Topics in Mining, Metallurgy and Materials Engineering

Series editor

Carlos P. Bergmann, Porto Alegre, Brazil

"Topics in Mining, Metallurgy and Materials Engineering" welcomes manuscripts in these three main focus areas: Extractive Metallurgy/Mineral Technology; Manufacturing Processes, and Materials Science and Technology. Manuscripts should present scientific solutions for technological problems. The three focus areas have a vertically lined multidisciplinarity, starting from mineral assets, their extraction and processing, their transformation into materials useful for the society, and their interaction with the environment.

More information about this series at http://www.springer.com/series/11054

Ivan Shabalov · Yury Matrosov
Alexey Kholodnyi · Maxim Matrosov
Valery Velikodnev

Pipeline Steels for Sour Service

 Springer

Ivan Shabalov
Association of Pipe Manufacturers
Moscow, Russia

Yury Matrosov
I. P. Bardin Central Research
    Institute for Ferrous Metallurgy
Moscow, Russia

Alexey Kholodnyi
I. P. Bardin Central Research
    Institute for Ferrous Metallurgy
Moscow, Russia

Maxim Matrosov
I. P. Bardin Central Research
    Institute for Ferrous Metallurgy
Moscow, Russia

Valery Velikodnev
LLC "Center for Examination
    of Pipeline Systems
    and Engineering"
Moscow, Russia

ISSN 2364-3293   ISSN 2364-3307 (electronic)
Topics in Mining, Metallurgy and Materials Engineering
ISBN 978-3-030-13128-9   ISBN 978-3-030-00647-1 (eBook)
https://doi.org/10.1007/978-3-030-00647-1

This Springer imprint is published by the registered company Springer Nature Switzerland AG
The registered company address is: Gewerbestrasse 11, 6330 Cham, Switzerland

Foreword

The experimental results and literature data on the technology development for the manufacture of steels for pipes intended for the transport of H_2S-containing oil and gas are summarized.

A contemporary view of the cracking mechanisms and the factors affecting the fracture resistance of low-alloy pipe steels in H_2S-containing media is presented.

The methods improving the quality of continuously cast slabs and the effect of chemical composition on the microstructure and properties of rolled pipe steels are considered. Much attention is paid to the physicometallurgical aspects of the microstructure formation and to the enhancement of the mechanical properties and HIC resistance the plates produced by thermomechanical rolling with accelerated cooling.

The book is intended for engineers and researchers of metallurgical enterprises and research institutes, for university professors, as well as for undergraduate and graduate students of relevant specialties.

Moscow, Russia   Georgy Filippov

Preface

In recent years, the number of developed gas and oil fields with an increased content of hydrogen sulfide grows, while previously such fields were considered to be of little use for operation. This is explained by the depletion of "clean" fields, which have been the main source of hydrocarbons for many decades. At the same time, there is a growing demand for gas–oil pipes, which are much more cracking resistant in an aggressive hydrogen sulfide-containing medium, the so-called sour gas, than the steels used to manufacture pipes that transport non-aggressive natural gas or oil. To prevent damage of pipelines transmitting "sour gas" from fields to plants for its purification from harmful H_2S, CO_2, etc., impurities, special requirements have been developed for the resistance of pipe metal to failure in hydrogen sulfide-containing media. The fulfillment of such requirements is necessary to ensure a lifelong operation of the pipes.

For decades, the metallurgical industry and physical metallurgy have solved the problem of creating steels with high resistance to various types of failure in sour environments such as hydrogen-induced cracking and sulfide stress cracking. A substantial progress was achieved, which led to the development of special steels and technologies for their production in all areas, from steelmaking and continuous casting to thermomechanical processing of rolled steel and pipe manufacture.

The book presents a contemporary view of the physicometallurgical factors affecting the fracture resistance of low-carbon low-alloy pipe steels in hydrogen sulfide-containing media. The methods for estimating the resistance of steel to hydrogen-induced cracking and the requirements for sour-gas-resistant electric-welded pipes are considered.

It is shown that, to improve the cracking resistance of steel under the effect of H_2S-containing medium, it is necessary to eliminate the causes promoting the nucleation and propagation of cracks. In addition to the cleanness from harmful impurities and non-metallic inclusions, the microstructure, in particular, in the central segregation zone of rolled plates substantially affects the resistance of steel to hydrogen-induced cracking. Therefore, special attention is paid in the book to the minimization of the negative effect of axial segregation on the hydrogen-induced cracking resistance of steel. Methods for improving the quality of continuously cast slab relative to the axial chemical heterogeneity as well as the effect of chemical composition on the microstructure and properties of high-quality flat rolled products from low-alloy pipe steels are considered.

A modern way to control the structure formation in steel is the combined process of controlled rolling and accelerated cooling. The physicometallurgical aspects of the microstructure formation and increase in mechanical properties and hydrogen-induced cracking resistance of the plates manufactured by thermome-chanical treatment by the scheme "controlled rolling followed by accelerated cooling" are considered.

A vision is given of the current level of technology for the manufacture of rolled steel for sour gas pipes at a number of leading Russian and foreign metallurgical enterprises. At present, mainly pipes of grades up to X65 inclusive have found commercial use for the construction of sour-gas-resistant pipelines. Research works aimed at the development of X70 and X80 grade pipes for operation in the most aggressive media are underway.

The materials of the book are based on our own research in laboratory and industrial conditions and also include an overview of a huge volume of Russian and world literature.

Moscow, Russia

Ivan Shabalov
Yury Matrosov
Alexey Kholodnyi
Maxim Matrosov
Valery Velikodnev

Contents

Chapter 1
Effect of Hydrogen Sulfide-Containing Media on Pipe Steels

Consequences of simultaneous and delayed failure of the pipe walls, equipment, and metal parts operated in oil and gas production areas in contact with liquid and gaseous media containing hydrogen sulfide have led to the need to study the mechanisms of the failure initiation. An artificial increase in the hydrogen concentration in steel allowed the observation of the fracture and embrittlement phenomena such as those occurring upon contact with a hydrogen sulfide-containing medium. This gave grounds to believe that the presence of hydrogen sulfide impurities in the service environment is the cause of the observed steel damage initiated by the presence of hydrogen in the steel. Further studies were held in the direction of eliminating the negative effect of hydrogen.

The chapter presents a contemporary view of the mechanisms of low-alloy pipe steels fracture in hydrogen sulfide-containing media. The methods for estimating the resistance of steel to hydrogen-induced cracking and the requirements for sour-gas-resistant electric-welded pipes are considered.

1.1 Mechanisms of Steel Fracture in H_2S-Containing Media

Hydrogen sulfide increases the corrosion rate of the metal equipment by a factor of tens compared to that in the media free from its compounds. Upon contact with moisture, hydrogen sulfide contained in gas and oil dissolves in it and dissociates to form H^+ ions, thus reducing the pH and forming an acidic corrosive medium, so-called sour gas. The dissociation of hydrogen sulfide occurs stepwise, mainly by the first step:

$$H_2S \rightleftarrows HS^- + H^+ \,(step\,I),$$

$$HS^- \rightleftarrows S^{2-} + H^+ \,(step\,II).$$

© Springer Nature Switzerland AG 2019
I. Shabalov et al., *Pipeline Steels for Sour Service*, Topics in Mining, Metallurgy and Materials Engineering, https://doi.org/10.1007/978-3-030-00647-1_1

The acidic medium enters an electrochemical reaction with the steel and causes an anodic reaction such as ionization of iron, which is described by the following equations:

$$Fe + H_2S + H_2O \rightarrow Fe(HS^-)_{ads} + H_3O^+,$$
$$Fe(HS^-)_{ads} \rightarrow (FeHS)^+ + 2e,$$
$$(FeHS)^+ + H_3O^+ \rightarrow Fe^{2+} + H_2S + H_2O.$$

With the formation of the $Fe(HS^-)$ chemisorption catalyst, which is adsorbed on the metal surface, the bond between the iron atoms is weakened, which facilitates their ionization.

The mechanism of the action of hydrogen sulfide on the cathodic reaction is as follows:

$$Fe + HS^- \rightarrow Fe(HS^-)_{ads},$$
$$Fe(HS^-)_{ads} + H_3O^+ \rightarrow Fe(H-S-H)_{ads} + H_2O,$$
$$Fe(H-S-H)_{ads} + e \rightarrow Fe(HS^-)_{ads} + H_{ads}.$$

Hydrogen sulfide does not participate directly in the cathodic reaction, but plays the role of a catalyst that accelerates the discharge of hydrogen ions. As a result of the cathodic reaction, atomic hydrogen is formed on the surface of the material. Iron sulfides are also formed. They are mainly represented as a sulfide film consisting of troilite (FeS) and pyrite (FeS_2). CO_2 contained in the environment also dissolves in water and reduces the pH of the service medium, which becomes more acidic.

The reduced hydrogen atoms are attracted to the metal surface and adsorbed by it due to the presence of free electrons. Upon the contact with the metal surface, the atomic hydrogen can pass to the H_2 molecular form or can be absorbed by (penetrate into) the metal. The absorption of hydrogen by the metal is facilitated by the presence of sulfide ions, which hinder the recombination of hydrogen into the molecular form.

The hydrogen concentration in the steel (Per_{Fe}) depends on the factors of the hydrogen sulfide-containing medium, such as the H_2S partial pressure (P_{H_2S}) and the pH (Fig. 1.1). The hydrogen absorption by steel increases with increasing H_2S partial pressure or with decreasing pH. An increase in gas pressure affects the intensity of corrosion processes, increasing the H_2S concentration per unit volume and the H_2S solubility in the aqueous phase.

Figure 1.2 shows the regions of the aggressiveness of hydrogen sulfide-containing media with respect to the cracking of low-alloy steels. When operating in a medium with a P_{H_2S} of <0.3 kPa (region 0), no precautions are generally required for steel choice. Nevertheless, it is necessary to take into account a variety of factors that can affect the behavior of steel in a given environment. The medium, in which the possibility of cracking occurs, is a solution with a hydrogen sulfide partial pressure of >0.3 kPa (regions 1–3). The medium aggressiveness increases with decreasing

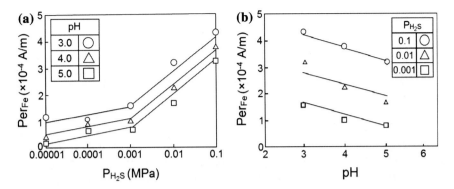

Fig. 1.1 Effect of the hydrogen sulfide partial pressure P_{H_2S} (**a**) and the medium acidity pH (**b**) on the parameter of hydrogen penetration into steel (Per_{Fe}) (Shabalov et al. 2017)

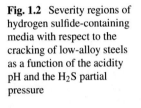

Fig. 1.2 Severity regions of hydrogen sulfide-containing media with respect to the cracking of low-alloy steels as a function of the acidity pH and the H_2S partial pressure

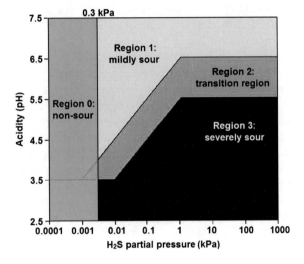

pH and/or increasing H_2S partial pressure. In such environments, one should use the steels highly resistant to corrosion cracking.

The hydrogen atoms, which have diffused into the metal volume, are distributed in a certain way among the metal atoms. The diffusivity of hydrogen in metal is higher by several orders of magnitude than those of other elements contained in the alloy. For example, for 100 h, the hydrogen atom in ferrite can pass a path 2 cm long, while typical interstitial impurities (C, N) pass only 10^{-6} mm, and the substitutional atoms remain virtually immobile. Hydrogen atoms in metal act as interstitial impurity. They are present in the crystal lattice between iron atoms and alloying elements or are accumulated in "traps" such as interfaces between non-metallic inclusions and matrix and other defects in the metal structure (grain boundaries and interphases, pores, etc.). Atomic hydrogen can pass into the molecular form of H_2 in the regions of the heterogeneity of the steel matrix. Since the resulting hydrogen gas occupies a substantial volume, the process of hydrogen transition from the atomic state to the molecular one leads to the excessive internal pressure, to high internal tensile stresses and, in critical cases, to the formation of microcracks. After the nucleation stage, the hydrogen-initiated crack propagates by the mechanism of the coalescence of several pores. The resulting cracks propagate mainly through hard and brittle microstructure constituents.

The interaction between steel and hydrogen leads to *"hydrogen embrittlement "*. The character of its manifestation is related to the nature of the phases formed upon such interaction. The term "hydrogen embrittlement" is conventional, since hydrogen often does not lead to purely brittle fracture. Decrease in plasticity can fluctuate in a wide range, from a few percents to a total loss of plasticity. The term *"hydrogen brittleness"* is understood to mean the whole set of negative phenomena caused by increased hydrogen content in metal.

The effect of hydrogen on the mechanical properties of metals can be realized by facilitating the ductile "cup-and-cone" fracture, which is customary for a given material, or by changing the character of fracture under the action of hydrogen from "normal" ductile fracture including the nucleation and growth of pores to low-ductile intra- and intercrystalline cleavage fracture.

The nature of hydrogen brittleness of metals is determined by the hydrogen content, the character of metal–hydrogen interaction, the state of hydrogen in the metal, the magnitude of the acting stresses (external and internal), the stressed state scheme, etc. Hydrogen can affect the nucleation of cracks, their propagation, or both the stages simultaneously. Because of such wide variety of factors, there is no single mechanism of hydrogen brittleness of metals, and even for the same metal, the operating mechanism of hydrogen embrittlement changes when the factors listed above are changed. The probability of simultaneous action of several mechanisms is high.

Hydrogen in the atomic and/or molecular state in steel can lead to the fracture of steel mainly by two mechanisms (Fig. 1.3): cracking occurring without stress and cracking occurring upon the action of residual and/or applied stress.

Cracking that occurs without stress manifests itself in the form of cracks located in the planes parallel to the surface of the rolled metal or pipe and is designated by the following terms (Fig. 1.4) (ISO 15156-2 2017):

- *hydrogen-induced cracking (HIC),* which is the lamellar plate cracking caused by the diffusion of atomic hydrogen, its accumulation in traps, and the formation of molecular hydrogen. Cracking occurs due to a sharp increase in pressure as a result of the recombination of hydrogen from the atomic to the molecular state. Such cracks appear in the regions with a high density of non-metallic inclusions, especially of flat shape, and the areas of anomalous structure (e.g., banded structure) resulting from the segregation of impurities and alloying elements in steel;
- *stepwise cracking (SWC),* the type of hydrogen-induced cracking, at which cracks in adjacent planes of the steel structure join and form "steps." The interconnection of hydrogen cracks with the formation of stepwise cracking depends on the level of local deformation between adjacent cracks and the brittleness of the matrix. HIC and SWC are usually characteristic of plate steel of low strength;

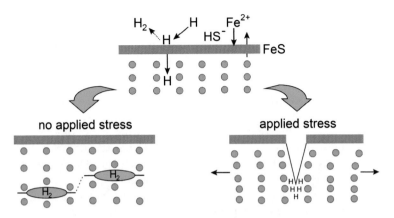

Fig. 1.3 Schematic representation of the mechanisms of steel cracking under the effect of absorbed hydrogen in a hydrogen sulfide-containing medium

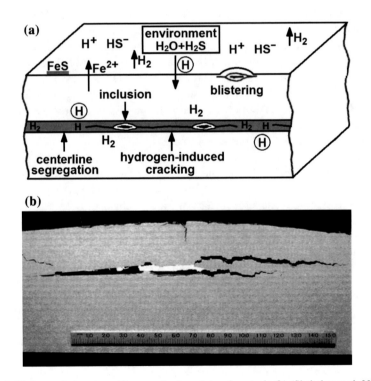

Fig. 1.4 Hydrogen-induced cracking: mechanisms (**a**) and example (**b**) (Shabalov et al. 2017)

- *blistering cracking* is the hydrogen cracking in subsurface layers of steel, and it manifests itself as a blistering on the steel surface.

The types of cracking occurring under the action of residual (internal) and/or applied (external) stresses are designated by the following terms (ISO 15156-2 2017):

- *sulfide stress cracking (SSC)* is the cracking of metal under the action of tensile stresses in the sour medium. Sulfide stress cracking is associated with the embrittlement of the metal by atomic hydrogen. Atomic hydrogen decreases the plasticity of steel and increases its susceptibility to cracking. In the case of steel failure by the SSC mechanism, the crack formation caused by the pressure of molecular hydrogen should also be taken into account. Such failure is typical of high-strength metal materials and the zones of increased hardness of welded joints;
- *soft zone cracking (SZC)* is a form of SSC that can occur in material if there are local regions of low yield strength ("soft zones"). Under operational loads, soft zones can be deformed and locally accumulate plastic deformation, thus increasing the sensitivity to SSC of the material, which is resistant to SSC in other cases. Such soft zones are usually associated with weld joints;
- *stress-oriented hydrogen-induced cracking (SOHIC)* is represented by fine cracks formed in the direction perpendicular to the stress; this leads to the arrangement of cracks into "stairs" connecting the preexisting (sometimes fine) hydrogen cracks. This form of cracking can be attributed to SSC caused by the combination of stress and local strain around the hydrogen cracks. SOHIC is associated with both SSC and HIC. It is observed in the base material of the longitudinal-welded pipes and in the heat-affected zone (HAZ) of welds in high-pressure vessels. SOHIC is a relatively rare phenomenon, which is generally associated with low-strength ferritic steels for pipes and pressure vessels;
- *hydrogen stress cracking (HSC)* is the cracking in metals that are not sensitive to SSC, but can be embrittled under the effect of hydrogen upon galvanic interaction as a cathode with another metal actively corroding as an anode. The term "galvanic HSC" (GHSC) is used just for such a cracking mechanism;
- *stress corrosion cracking (SCC)* is the cracking of metal under the effect of anodic processes of local corrosion and tensile stress in the presence of water and hydrogen sulfide. Chlorides and/or oxidants at elevated temperatures can increase the susceptibility of metals to such corrosion mechanism.

From the above types of stress-induced cracking, SSC, SOHIC (Fig. 1.5), and SZC are characteristic of the operating conditions of low-alloy pipe steels.

The amount of hydrogen entering the metal upon the interaction with the hydrogen sulfide-containing medium is much higher than the content of metallurgical hydrogen present in the steel in the initial state. At the modern steelmaking and continuous casting technology, the content of hydrogen in liquid steel is about 2–3 ppm, which in combination with anti-flake treatment of slabs and plates allows one to substantially reduce or eliminate the occurrence of failure cracks, flakes, and other defects in steel billets and rolled products.

(a)

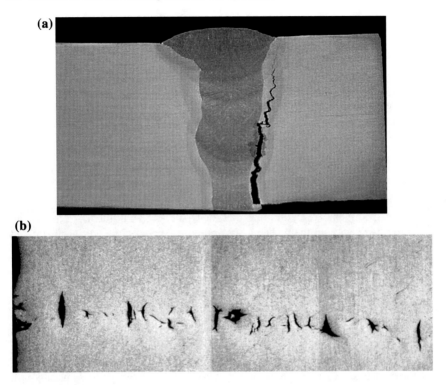

(b)

Fig. 1.5 Types of cracks occurring under the action of stress: SSC (**a**) and SOHIC (**b**) (Shabalov et al. 2017)

1.2   Methods of Tests for Cracking Resistance in Hydrogen Sulfide-Containing Media

The methods of test and evaluation of steels for cracking resistance in hydrogen sulfide-containing media have been improved for a long time. At present, tests for the cracking resistance of steels in hydrogen sulfide-containing media are carried out by the action of an aggressive medium with or without load applied to specimen, assembly, or structure with subsequent monitoring of cracking. Such tests are performed for the determination of the causes, character, and mechanisms of fracture, for the substantiation of the choice of the material used, and for the assessment and diagnosis of materials operating in aggressive conditions. In most cases, the test results are evaluated by using the parameters, which are compared for the determination of the resistance to one or another type of cracking of the test material. The criteria for cracking resistance should characterize the properties of the material and take the same values at different methods of determination. This allows one to use such criteria for the classification and standardization of steels and test methods.

The tests are performed in production or operation environments or in the media for accelerated testing. The methods can be divided into laboratory, bench, and field tests. The development of fracture by the mechanism of corrosion cracking of metal upon the pipe operation can be a long process. Therefore, when performing the tests, it is important to have the methods available that allow a relatively short-term evaluation of the long-term operation behavior of the material. Acceleration of the development of cracking processes identical to those occurring upon operation is achieved by the stimulation of reactions by aggressive components, increasing stress, temperature elevation, etc.

Numerous methods of corrosion testing have the following identical elements: shape and dimensions of the specimens, medium and load parameters, conditions for the test termination, post-expositional evaluation, and criteria for cracking resistance. The specimens are cylindrical, flat, or U-shaped, with notches, cracks, and concentrators or without them. The loads applied to the specimen differ in stressed states, the characters of the changes in time, and the application methods. Tensile stresses upon uniaxial tension or bending are used most often. The load is applied either permanently or cyclically. If the cracking mechanism is not related to an external stress, the tests are carried out on unloaded specimens.

The metal is hydrogenated by the action of the medium simulating the production conditions. As a rule, hydrogen sulfide is introduced into the water-based solution up to the saturation state (2.2–3.5 g/l) at atmospheric pressure and at a temperature of 25 ± 5 °C. For more probable fracture, the pH of the solution is reduced by acidifying it with ethane, acetic, and sulfuric acids; sometimes, the solution is saturated with carbon dioxide.

The test is terminated upon reaching either the specified duration or the limit state. In the latter case, the reasons for the test termination are the loss of the load-carrying capacity of the specimen because of the corrosion crack propagation, the apparent failure of the specimen, or the achievement of a certain critical size by corrosion cracks. Upon the tests for a predetermined duration, the change in the service properties of the metal is estimated after exposure for the test time.

The effectiveness of laboratory testing methods is largely determined by the comparability and reproducibility of the results and the degree, to which they correspond to the actual operating conditions. The cracking resistance of steel in hydrogen sulfide-containing media is determined in specialized laboratories of corrosion tests.

The main international standard laboratory test methods were developed by NACE (National Association of Corrosion Engineers, headquartered in Houston, USA):

- ANSI/NACE Standard TM0284-2011. Standard Test Method "Evaluation of Pipeline and Pressure Vessel Steels for Resistance to Hydrogen-Induced Cracking";
- ANSI/NACE Standard TM0177-2016. Standard Test Method "Laboratory Testing of Metals for Resistance to Sulfide Stress Cracking and Stress Corrosion Cracking in H_2S Environments."

The methodological features of testing and evaluation of the resistance of pipe steels to cracking in hydrogen sulfide-containing media are considered below.

Test for Resistance to Hydrogen-Induced Cracking (HIC, SWC, Blistering)
The NACE TM0284 standard establishes the test method for evaluating pipeline and pressure vessel steels for resistance to hydrogen-induced cracking caused by hydrogen absorption resulting from corrosion in a wet H_2S-containing medium. The standard was developed in 1984 and in the initial version described a mandatory set of test conditions for pipe steels operating in hydrogen sulfide-containing media. The test procedure was extended to plates for high-pressure vessels in 1996 and to steel fittings and flanges in 2011.

The test method consists in holding the specimens in a hydrogen sulfide-containing solution without applying external stress at ambient temperature and pressure and their post-exposure assessment. The test media used are standard solutions saturated with hydrogen sulfide:

- test Solution A (the so-called NACE solution), which is the solution of 5% sodium chloride (NaCl) and 0.5% glacial acetic acid (CH_3COOH) in distilled or deionized water;
- test solution B (the so-called BP solution), which is an artificial seawater solution according to ASTM D1141.

Figure 1.6 shows the schemes of cutting test specimens from plates and longitudinally welded pipes. Three specimens are tested. The pieces for making specimens from the plates are taken from one end along the middle of the plate width with the longitudinal axis parallel to the main rolling direction (see Fig. 1.6a). Specimens from the body of the pipe (90° and 180° from the seam) are taken in the direction of the longitudinal pipe axis (see Fig. 1.6b), and specimens from the weld seam are taken perpendicular to the seam (see Fig. 1.6c). Specimens are cut from the workpieces 100 ± 1 mm long and 20 ± 1 mm wide. The thickness of the specimen for testing the plates and the pipe body is equal to the total thickness of the plate or pipe wall. A layer not more than 1 mm deep can be removed from the surfaces of the workpieces. The pieces for the test specimens from the pipes should not be straightened.

After preparation, the specimens are put into a sealed vessel (autoclave), where they are isolated from the vessel and other test specimens by glass or other non-metallic rods of at least 6 mm in diameter. Then, the vessel is filled with a model medium in such a way that the ratio of the solution volume to the total surface area of the specimen is at least 3 ml/cm². The solution in the autoclave is purified from air with nitrogen for at least 1 h. After purification, H_2S is passed through the solution. The rate of hydrogen sulfide bubbling for the first 60 min should be at least 200 cm³/min per 1 L of solution. Then, an increased H_2S gas pressure is maintained at a constant gas flow rate to provide the saturation of the solution. The H_2S concentration in the solution should be at least 2.3 ppm (mg/m³) at the end of the test. If Solution A is used, the pH should be maintained between 2.6 and 2.8 at the beginning of the test. For Solution B, the pH should be 8.1–8.3.

The test duration is 96 h. The solution temperature during the test is 25 ± 3 °C. At the end of the test, the pH of Solution A should not exceed 4.0, and the pH of Solution B is between 4.8 and 5.4.

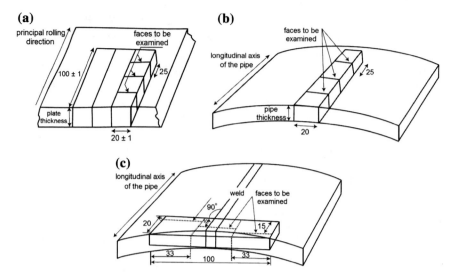

Fig. 1.6 Scheme of cutting the test specimens from plates for pipes (**a**) and from longitudinal-welded pipes (**b, c**): base metal (**b**) and weld seam (**c**) (ANSI/NACE TM0284 2011)

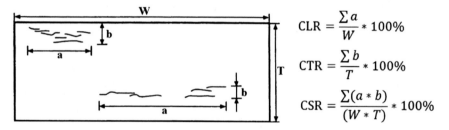

$$CLR = \frac{\sum a}{W} * 100\%$$

$$CTR = \frac{\sum b}{T} * 100\%$$

$$CSR = \frac{\sum (a * b)}{(W * T)} * 100\%$$

Fig. 1.7 Estimated surface of the HIC specimen cross section, the crack parameters, and the formulas for calculating the CLR, CTR, and CSR parameters, where W is the specimen width, T is the specimen thickness, a is the crack length, and b is the crack thickness (ANSI/NACE TM0284 2011)

After completion of the test, the H_2S feed is stopped, and the solution is purged with nitrogen to remove hydrogen sulfide. The specimens are taken and cleaned to remove scale and corrosion deposits. Then, the specimens are cut for the study according to the scheme shown in Fig. 1.6 and subjected to metallographic grinding and etching to enhance the visibility of possible cracks.

After preparation of the surface, the cross sections of the specimens are inspected for cracks in a microscope. In case where cracks are detected, they are measured for their sizes and relative positions (Fig. 1.7). Based on measurements of crack geometric parameters for each cross section, the following HIC parameters are calculated by formulas: the *crack length ratio (CLR)*, the *crack thickness ratio (CTR)*, and the *crack sensitivity ratio (CSR)*. For each test specimen, the average value over three sections is determined.

In addition to the standardized method of metallographic evaluation of hydrogen cracking criteria, the ultrasonic inspection method (Fig. 1.8) is used to determine the *crack area ratio (CAR)* (Bosch et al. 2008). Unlike CLR, CTR, and CSR, the CAR value is not an absolute criterion, since it can depend on the applied ultrasound method, instrument calibration, etc. An example of hydrogen cracks in the axial zone and on the surface of the plate is shown in Fig. 1.9.

Test for Resistance to Sulfide Stress Cracking (SSC) The NACE TM0177 standard establishes laboratory test methods for evaluating the resistance of metals to fracture by cracking under the combined action of tensile stress and corrosion in hydrogen sulfide-containing aqueous media with a low pH. The standard provides tests for the resistance of metals to sulfide stress cracking (SSC) and stress corrosion cracking (SCC, HSC).

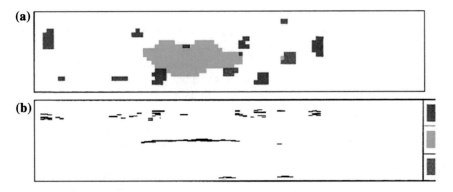

Fig. 1.8 Estimation of the HIC specimen by ultrasonic inspection method: **a** frontal projection and **b** lateral projection (Bosch et al. 2008)

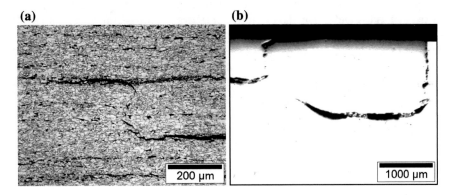

Fig. 1.9 Hydrogen-induced cracking in the centerline segregation zone (**a**) and on the surface (**b**) of the plate (Bosch et al. 2008)

The following test methods are provided in the standard:

- method A—NACE Standard Tensile Test;
- method B—NACE Standard Bent Beam Test;
- method C—NACE Standard C-Ring Test;
- method D—NACE Standard Double Cantilever Beam (DCB) Test.

Additionally, the references are indicated to the test method for four-point bending of the specimens.

Tests can be performed both under atmospheric conditions and at elevated temperatures and pressures. The occurrence of cracking by the SSC mechanism is associated with operation at room temperature, and SCC is related to higher temperatures. Carbon and low-alloy steels are tested at room temperature, at which the susceptibility to SSC is high.

It is noted in the standard that the specimens should be selected so that the test results are most representative and expectable under the corresponding operating conditions. All specimens in the series should be of the same orientation and similar in the microstructure and mechanical properties. To determine the properties of the metal, a tensile test is performed in accordance with accepted standard techniques. It is recommended to make the tensile test specimen and the SSC resistance test specimen from adjacent sections and at the same position and orientation. Anisotropy of mechanical properties and the cracking susceptibility of the material can play an important role. The direction of crack propagation in the specimen should correspond to the expected path in the actual construction. Considering the tensile hoop stresses in the pipes, it is expedient to take the specimens perpendicularly to the rolling direction of the plate, which corresponds to the longitudinal axis of the straight pipe.

The test medium can initiate the failure of steel by the mechanism of hydrogen-induced cracking. This is especially true for low-strength steels, which usually are not susceptible to SSC. Hydrogen-induced cracking can be detected by visual inspection and metallographic control.

For the pipe steels under consideration, Solution A similar to that presented in NACE TM0284 is used as the model medium for the SSC test. For low-alloy pipe steels, method A and the method of four-point bending of specimens at room temperature and atmospheric pressure are most often used. The methods are considered below.

Method A is a standard tensile test used to evaluate the cracking resistance under the action of a wet hydrogen sulfide-containing medium under uniaxial tensile loading of specimens. The specimens loaded to a certain stress level either fail or do not fail for a given time. The susceptibility to SSC is usually determined by this method as the time to failure, i.e., to complete separation of two halves of the specimen. When testing a series of specimens at different loading levels, it is possible to establish an arbitrary threshold stress (σ_{thr}), at which SSC occurs.

Figure 1.10 shows the scheme and dimensions of the specimens used for the NACE tensile test.

Tests are carried out with constant load devices or sustained load devices. The device should not cause torsional loads. Sustained load tests can be conducted with

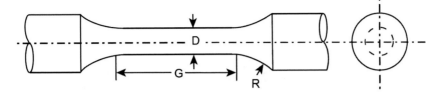

Dimension	Standard tensile test specimen	Subsize tensile test specimen
D (mm)	6.35 ± 0.13	3.81 ± 0.05
G (mm)	25.4	15
$R_{min.}$ (mm)	15	15

Fig. 1.10 Scheme and dimensions of the specimens used for the NACE tensile test (method A) (ANSI/NACE TM0177 2016)

(a)   **(b)**

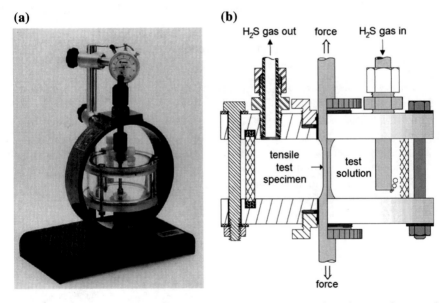

Fig. 1.11 Proof ring for sustained load tests (**a**) and tensile test specimen in an environmental chamber (**b**) (ANSI/NACE TM0177 2016)

spring-loaded devices and proof rings (Fig. 1.11). The test vessel size should be sufficient to maintain a solution volume of 20–40 ml per 1 cm^2 of the test specimen surface area.

After cleaning, the specimen is put into the test vessel and subjected to loading. The tensile load applied to the specimen is determined by the following equation:

$$P = k \cdot \sigma_{T\,spec} \cdot S$$

where

P	is the load acting on the specimen (N);
k	is the load factor;
$\sigma_{T\ spec}$	is the specified minimum yield strength (N/mm^2);
S	is the minimum cross-sectional area of the gage section of the specimen (mm^2).

The vessel is purged with an inert gas and then immediately filled with a test solution so that the gauge length of the specimen is completely immersed in the solution. The solution is saturated with H$_2$S at a flow rate of at least 100 ml/min for at least 20 min per 1 L of solution. The H$_2$S flow in the vessel and in the outlet trap should be continuous throughout the test at a low flow rate (a few bubbles per minute).

The test is terminated at test specimen failure or after 720 h if the specimen is not failed. Following exposure, the surfaces of the gauge section of the non-failed test specimens shall be cleaned and inspected for evidence of cracking. Warnings are:

- complete separation of the test specimen in less than 720 h;
- visual observation of cracks on the gauge section of the test specimen at 10× after completing the 720-h test duration.

At each stress level, the time to failure or the absence of failure with the presence of cracks at the end of the test is recorded.

Figure 1.12 shows the specimens subjected to the stress sulfide cracking (SSC) test under uniaxial tension.

In the presence of cracks on the gage surface of the specimen, they are examined to determine their cause. If it is confirmed that the cause of cracks is not SSC, then the specimen is considered to have passed the test. In some cases, defects perpendicular to the action of the applied load can be observed (Fig. 1.13). It can be seen that the defect does not extend into the interior of the specimen body in the form of a crack caused by SSC. Figure 1.14 shows a typical SSC nucleated on the specimen surface.

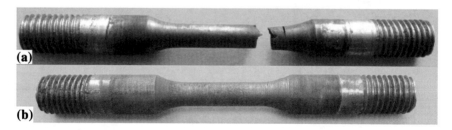

Fig. 1.12 Appearance of the specimens after the sulfide stress cracking (SSC) resistance test upon uniaxial tension: **a** specimen failed in 48 h and **b** specimen without failure after test for 720 h (Shabalov et al. 2017)

(a)

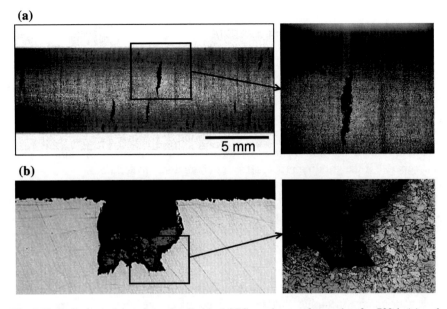

(b)

Fig. 1.13 Defects on the surface of cylindrical SSC specimens after testing for 720 h (**a**) and longitudinal section of the specimen at the location of the defect (**b**)

Fig. 1.14 SSC on the specimen surface, ×500

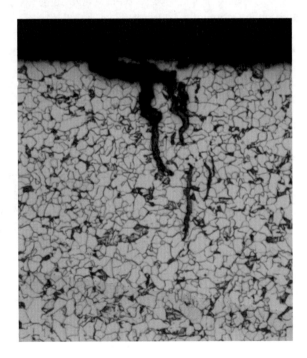

To prevent premature failure caused by defects on the surface of the specimens, the following additional requirements for the manufacture of specimens are recommended:

- eliminate undercutting of fillet radii in machined specimens;
- machine the test specimen gauge section with a slight (0.05–0.13 mm) taper that produces a minimum cross section in the middle of the gauge section;
- the test specimen must be machined or ground carefully to avoid overheating and cold working in the gauge section. In machining operations, the final two passes should remove no more than a total of 0.05 mm of material. Grinding is also acceptable if the grinding process does not harden the material;
- final surface finish may be obtained by mechanical polishing or electropolishing and shall be 0.81 μm or finer;
- before testing, test specimens should be degreased with solvent and rinsed with acetone. The gauge section of the test specimen should not be handled or contaminated after cleaning.

Figure 1.15 shows an example of the specimen failure at the edge of the gauge zone near the fillet. Probable cause of such failure can be the transverse bands left after poor-quality specimen manufacture. Such bands are stress concentrators and reduce the SSC resistance of the steel. Therefore, upon the manufacture of specimens for SSC testing, it is necessary to make high demands on the surface quality of both the gauge section of the specimen and the fillets.

For more accurate determination of stress σ_{thr} that does not cause failure upon the SSC test, additional specimens are tested at different loads, and, according to the test results, serial curves are plotted (Fig. 1.16).

The SSC resistance test of steel upon four-point bending of specimens was carried out in accordance with NACE TM0177 technique in test Solution A. The tests are carried out on three specimens 115 mm long, 15 mm wide, and 5 mm thick according to ISO 7539-2 or ASTM G39.

Fig. 1.15 Surface of the specimen near the fillet; failure occurs at the edge of the gauge zone with traces of cutting

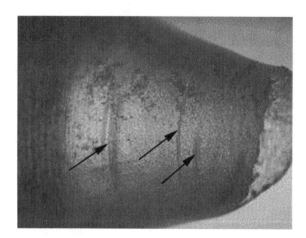

Fig. 1.16 Exposure time of SSC specimens from X60 grade steel plates 20 mm thick at different load levels from the nominal yield strength (425 N/mm^2)

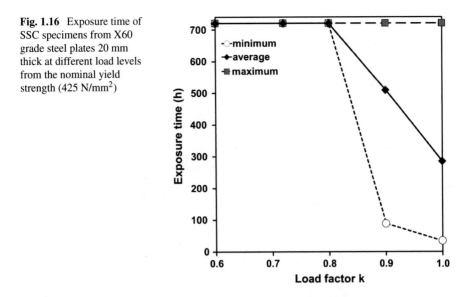

Before testing, the specimens are subjected to static loading by bending at four points in the clamping device (Fig. 1.17). The bending of the specimen is given in millimeters of the deviation from its original axis.

After loading in the clamping device, the specimen is put into a test vessel and the test is performed in a manner similar to the HIC test. The exposure time is 720 h. After the completion of the test, the specimens are removed from the medium and the clamping device and the tensile surface is evaluated for cracks in an optical microscope at a magnification of ×10 (see Fig. 1.18).

The specimen is considered as passed the test if no cracks (see Figs. 1.18a and 1.19b) or signs of failure (see Fig. 1.19a, c) are found on the tensile surface.

Fig. 1.17 Bending test specimen in a four-point clamping device (Shabalov et al. 2017)

(a)

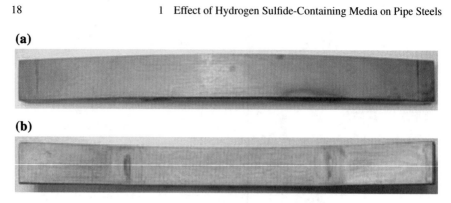

(b)

Fig. 1.18 Surface of the specimen after four-point bending test: **a** bent surface (subjected to tension) and **b** internal surface (subjected to compression) (Haase et al. 2012)

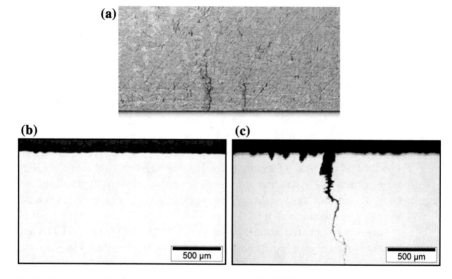

Fig. 1.19 Surface subjected to tension, the specimen with SSC (**a**), longitudinal plane of the specimen without crack (**b**) and with SSC (**c**) (Haase et al. 2012)

1.3   Requirements for Pipes Ordered for Sour Service

The equipment of the gas and oil industry is operated under extremely severe conditions, and the durability and reliability of its operation largely depend on the technical characteristics of the materials used. They should have a certain combination of strength and plasticity, cold resistance, and other technological properties, and should not be scarce and expensive. Many oil and gas fields are located in remote and hard-to-reach areas; therefore, the number of producing fields with an increased content of hydrogen sulfide impurities increases. Stringent requirements are imposed on such steels for corrosion resistance, especially for specific failure types, namely hydrogen-induced cracking, sulfide stress cracking, general corrosion. Such pipelines are used to transport hydrogen sulfide-containing hydrocarbons at production sites and from fields to processing plants and other large consumers of hydrogen sulfide-containing gas.

The "instructions for the selection and use of materials for the manufacture of pipes for pipelines operating in hydrogen sulfide-containing media" developed by VNIIGAZ (LLC «VNIIGAZ» 2000) presents: the graduation of corrosive hydrogen sulfide-containing media according to their aggressiveness with respect to pipelines in the gas industry; requirements for pipes used for the construction of pipelines transporting hydrogen sulfide-containing media; recommended selection of pipes for the conditions with a different aggressiveness of working environments and for pipelines with different degrees of responsibility (categories); and calculation of wall thickness of pipes with allowance for the category of the section and the degree of corrosive aggressiveness of the working environment.

When choosing pipes for the arrangement and reconstruction of gas and oil production, transportation, and processing facilities containing hydrogen sulfide and associated components (condensate, carbon dioxide, stratal water impurities, etc.), it is often necessary to vary the choice of pipes with allowance for both the operating conditions of the specific pipeline (the degree of aggressiveness of the working environment, temperature, pressure, working stress, the degree of responsibility determined by the possible consequences of the failure of the pipeline or its section, etc.) and the properties of the pipes supplied by certain specifications, with certain guaranteed properties such as SSC and HIC resistance, strength, plasticity, cold resistance, weldability.

Hydrogen sulfide-containing corrosive media include gaseous, liquid, and gas–liquid mixtures, which, due to the presence of hydrogen sulfide and moisture, cause the failure of steel pipes and products as a result of electrochemical corrosion including general corrosion, SSC, and HIC. According to the ability to cause cracking of steels as well as the intensity of the impact on the environment upon entering the atmosphere, the hydrogen sulfide-containing media are classified as media with low, medium, and high hydrogen sulfide contents. The first media include those with a partial hydrogen sulfide pressure P_{H_2S} of 0.3–10 kPa (0.003–0.1 kgf/cm^2). The media with a medium hydrogen sulfide content include those with a P_{H_2S} of 10 kPa–1.5 MPa (0.1–15 kgf/cm^2). The media with a high content of hydrogen sulfide are characterized by a PP_{H_2S} of >1.5 MPa (15 kgf/cm^2). The media with a partial pressure P_{H_2S} of <0.3 kPa are considered as those unprovoking cracking.

For the construction of pipelines serving for the transportation of hydrogen sulfide-containing media, the pipes made of carbon and low-alloy high-quality killed steels should be used. The satisfactory SSC resistance of such steels should be established by laboratory tests and confirmed by testing in industrial field conditions including assembly welding or by industrial application at gas industry facilities with hydrogen sulfide-containing media. The pipes made of ordinary quality steel may be used in pipeline sections of low responsibility with low stresses from environmental pressure.

The API Spec 5L standard, the international standard ISO 3183-3, and the European standard EN 10208-2 are the basic normative documents regulating the general requirements for electric-welded pipes for gas and oil pipelines. The main Russian documents containing requirements for pipe steels and pipes of oil and gas assortment are SNiP 2.05.06-85 (SNiP 2.05.06-85 1985) and "instruction for the use of steel pipes in the gas and oil industry" approved by JSC Gazprom.

The international standards cover the pipe grades not intended for operation in H_2S-containing media, from B (L245) to X120 (L830). According to the SNiP 2.05.06-85 requirements, the strength range is extended to grades up to K65 (X80). The main international document for the design, construction, and operation of submarine pipelines is Det Norske Veritas (DNV) Offshore Standard OS F101.

API Spec 5L and ISO 3183-3 provide requirements for pipes of grades up to X70 (L485MS) operating in hydrogen sulfide-containing media.

To harmonize the international and Russian regulatory and technical documentation, the STO Gazprom specification 2-4.1-223-2008 "technical requirements for electrically welded hydrogen sulfide-resistant pipes" was issued, which contains a gradation of strength classes from K48 to K52 and from X42 to X52 (STO Gazprom 2008). The standard is applied to steel electrowelded pipes 530-1020 mm inclusive in diameter, single-seam longitudinal joints welded by submerged arc welding, which are designed for construction and repair of pipelines transporting hydrogen sulfide-containing media operating at a pressure of 5.4–9.8 MPa inclusive. The main technical requirements for assortment, geometric dimensions, chemical composition, properties, acceptance rules and test methods, marking, packing, storage, and transportation of pipes for various operating conditions are specified.

The following requirements are imposed on rolled plates and pipes, depending on the parameters, purpose, and criticality of the pipelines: geometric dimensions, production technology, chemical composition, mechanical properties, content of nonmetallic inclusions, grain size, structure banding, continuity detected by ultrasonic testing, and a number of other special requirements.

Steel for pipes intended for transporting H_2S-containing media should be produced in electric steelmaking units or by a converter process followed by the treatment (secondary treatment, vacuum degassing), which ensures the required cleanness from gases, harmful impurities, and non-metallic inclusions. Steel should be completely killed. Upon the out-of-furnace treatment, the steel should be calcium modified to control the shape of non-metallic inclusions.

Steel should be cast using the method of continuous casting. Upon casting, measures should be taken to reduce the contamination of steel with non-metallic inclusions and to reduce central segregation. Macrostructure defects in the continuously cast billets are evaluated according to the Russian method by OST 14-4-73 "steel. A method for controlling the macrostructure of a cast billet (ingot) obtained by the continuous casting method" or according to the Mannesmann method "Mannesmann rating system for internal defects of continuously cast slabs".

Table 1.1 presents the requirements for the chemical composition of seamless and welded hydrogen sulfide-resistant pipes according to ISO 3183-3 and API Spec 5L. The letters N, Q, and M indicate the delivery status: after normalizing rolling, normalizing, or normalizing and tempering (N); after quenching and tempering (Q);

after thermomechanical rolling (M); and S indicates operating conditions in acid media.

For the steels for sour service, tightened requirements are imposed on the contents of chemical elements, in particular, the mass fractions of C, Mn, S, and P.

At carbon content $\leq 0.12\%$, the carbon equivalent should be calculated by the following equation:

$$CE_{P_{cm}} = C + \frac{Si}{30} + \frac{Mn + Cu + Cr}{20} + \frac{Ni}{60} + \frac{Mo}{15} + \frac{V}{10} + 5B.$$

In the steels containing more than 0.12% carbon, the carbon equivalent is calculated by the equation:

$$CE_{IIW} = C + \frac{Mn}{6} + \frac{Cr + Mo + V}{5} + \frac{Ni + Cu}{15}.$$

Table 1.2 represents the requirements for the chemical composition of steel for hydrogen sulfide-resistant pipes in accordance with STO Gazprom 2-4.1-223-2008.

Contamination of steel by non-metallic inclusions (NMI) is determined in finished rolled products in accordance with GOST 1778. Tightened requirements are imposed on the contamination by NMI for the hydrogen sulfide-resistant pipes steels (Table 1.3). In addition, there may be restrictions on contamination by spot oxides (SO), plastic silicates (PS), and nitrides (N).

The requirements for the mechanical properties of the base metal of the pipes upon the tensile test according to ISO 3183-3 and API Spec 5L are given in Table 1.4.

Requirements for the mechanical properties of the pipes according to the STO Gazprom 2-4.1-223-2008 are given in Table 1.5.

The requirements for mechanical properties and cold resistance of rolled plates are formulated by the pipe manufacturer, based on the need to allow for the effect of pipe processing (pipe dimensions, forming, welding, expansion, coating, etc.), and are often more stringent than the requirements imposed on pipes. The minimum yield strength and ultimate tensile strength are increased by 15–30 N/mm^2, while the maximum ones are reduced by 15–30 N/mm^2. The requirements on the relative elongation are increased by 2–4%, the shear area of the DWTT specimens is increased by 5–15%, and the impact toughness is increased by 50–100 J/cm^2.

The grain size of the structure should not exceed number 9 on scale 1 of GOST 5639. The structure banding according to GOST 5640 (row A) should not exceed number 2.

The hardness of the metal of the pipe body, the weld seam, and the heat-affected zone should not exceed 250 HV$_{10}$ or 22 HRC according to ISO 3183-3 and API Spec 5L and 220 HV$_{10}$ according to STO Gazprom 2-4.1-223-2008.

The requirements for hydrogen sulfide cracking resistance of pipes are related to special properties. The resistance to hydrogen-induced cracking (HIC) is determined under laboratory conditions using the NACE TM0284 procedure. The resistance to sulfide stress cracking SSC is determined using the NACE TM0177.

Table 1.1 Chemical composition for PSL-2 pipes ordered for sour service according to API Spec 5L (API Spec 5L 2012) and ISO 3183-3 (ISO 3183-3 2012)

Steel grade	Mass fraction, based on heat and product analyses (wt%), maximum								Carbon equivalent[a] (%), maximum	
	C^b	Si	Mn^b	P	S	V	Nb	Ti	Other[c, d]	CE_{IIW}	CE_{Pcm}
SMLS and welded pipes										
L245NS or BNS	0.14	0.40	1.35	0.020	0.003^e	f	f	0.04	g	0.36	0.19^h
L290NS or X42NS	0.14	0.40	1.35	0.020	0.003^e	0.05	0.05	0.04	–	0.36	0.19^h
L320NS or X46NS	0.14	0.40	1.40	0.020	0.003^e	0.07	0.05	0.04	g	0.38	0.20^h
L360NS or X52NS	0.16	0.45	1.65	0.020	0.003^e	0.10	0.05	0.04	g	0.43	0.22^h
L245QS or BQS	0.14	0.40	1.35	0.020	0.003^e	0.04	0.04	0.04	–	0.34	0.19^h
L290QS or X42QS	0.14	0.40	1.35	0.020	0.003^e	0.04	0.04	0.04	–	0.34	0.19^h
L320QS or X46QS	0.15	0.45	1.40	0.020	0.003^e	0.05	0.05	0.04	–	0.36	0.20^h
L360QS or X52QS	0.16	0.45	1.65	0.020	0.003^e	0.07	0.05	0.04	g	0.39	0.20^h
L390QS or X56QS	0.16	0.45	1.65	0.020	0.003^e	0.07	0.05	0.04	g	0.40	0.21^h

(continued)

Table 1.1 (continued)

Steel grade	Mass fraction, based on heat and product analyses (wt%), maximum									Carbon equivalent[a] (%), maximum	
	C[b]	Si	Mn[b]	P	S	V	Nb	Ti	Other[c, d]	CE_IIW	CE_Pcm
L415QS or X60QS	0.16	0.45	1.65	0.020	0.003[e]	0.08	0.05	0.04	g, i, k	0.41	0.22[h]
L450QS or X65QS	0.16	0.45	1.65	0.020	0.003[e]	0.09	0.05	0.06	g, i, k	0.42	0.22[h]
L485QS or X70QS	0.16	0.45	1.65	0.020	0.003[e]	0.09	0.05	0.06	g, i, k	0.42	0.22[h]
Welded pipe											
L245MS or BMS	0.10	0.40	1.25	0.020	0.002	0.04	0.04	0.04	–	–	0.19
L290MS or X42MS	0.10	0.40	1.25	0.020	0.002	0.04	0.04	0.04	–	–	0.19
L320MS or X46MS	0.10	0.45	1.35	0.020	0.002	0.05	0.05	0.04	–	–	0.20
L360MS or X52MS	0.10	0.45	1.45	0.020	0.002	0.05	0.06	0.04	–	–	0.20
L390MS or X56MS	0.10	0.45	1.45	0.020	0.002	0.06	0.08	0.04	g	–	0.21
L415MS or X60MS	0.10	0.45	1.45	0.020	0.002	0.08	0.08	0.06	g, i	–	0.21

(continued)

Table 1.1 (continued)

Steel grade	Mass fraction, based on heat and product analyses (wt%), maximum									Carbon equivalent[a] (%), maximum	
	C[b]	Si	Mn[b]	P	S	V	Nb	Ti	Other[c, d]	CE$_{IIW}$	CE$_{Pcm}$
L450MS or X65MS	0.10	0.45	1.60	0.020	0.002	0.10	0.08	0.06	g, i, j	–	0.22
L485MS or X70MS	0.10	0.45	1.60	0.020	0.002	0.10	0.08	0.06	g, i, j	–	0.22

[a]Based on product analysis, the CE$_{IIW}$ limits apply if C > 0.12% and the CE$_{Pcm}$ limits apply if C ≤ 0.12%

[b]For each reduction of 0.01% below the specified maximum for C, an increase of 0.05% above the specified maximum for Mn is permissible, up to a maximum increase of 0.20%

[c]Al$_{total}$ ≤ 0.060%; N ≤ 0.012%; Al/N ≥ 2:1 (not applicable to titanium-killed or titanium-treated steel); Cu ≤ 0.35% (if agreed, Cu ≤ 0.10%); Ni ≤ 0.30%; Cr ≤ 0.30%; Mo ≤ 0.15%; B ≤ 0.0005%

[d]For welded pipe where calcium is intentionally added, unless otherwise agreed, Ca/S ≥ 1.5 if S > 0.0015%. For SMLS and welded pipes, Ca ≤ 0.006%

[e]The maximum limit for S may be increased to ≤0.008% for SMLS pipe and, if agreed, to ≤0.006% for welded pipe. For such higher S levels in welded pipe, lower Ca/S ratios may be agreed

[f]Unless otherwise agreed, Nb + V ≤ 0.06%

[g]Nb + V + Ti ≤ 0.15%

[h]For SMLS pipe, the listed CE$_{Pcm}$ value may be increased by 0.03

[i]If agreed, Mo ≤ 0.35%

[j]If agreed, Cr ≤ 0.45%

[k]If agreed, Cr ≤ 0.45% and Ni ≤ 0.50%

Table 1.2 Chemical composition for pipe in accordance with STO Gazprom 2-4.1-223-2008 (STO Gazprom 2008)

Chemical composition (wt%)						
C	Si	Mn	P	V	N	Al
≤0.15	0.150–0.400	≤1.20	≤0.015	≤0.060	≤0.080	0.020–0.050

1. Microalloying of steel with elements not specified in the table is permitted for obtaining the required properties
2. Mass fractions of copper, nickel, and chromium should exceed 0.25% each, with a total fraction of not more than 0.90%
3. The mass fractions of calcium and molybdenum in the steel are allowed to be not more than 0.005 and 0.35%, respectively
4. No boron addition is allowed
5. Maximum sulfur contents for resistance groups
   0.002% for C-1
   0.003% for C-2
   0.005% for C-3

Table 1.3 Norms of steel contamination by non-metallic inclusions in accordance with STO Gazprom 2-4.1-223-2008 (STO Gazprom 2008)

Type of inclusion	Maximum number	Mean number
Sulfides (S)	1.5	1.0
Stitched oxides (SO)	2.5	2.0
Brittle silicates (BS)	2.5	2.0
Non-deformable silicates (NS)	2.5	2.0

The parameters of corrosion cracking resistance are as follows:

- crack length ratio (CLR);
- crack thickness ratio (CTR);
- crack sensitivity ratio (CSR);
- conditional threshold stress σ_{thr}, at which, for the basic test time (720 h), there is no failure upon testing cylindrical specimens for uniaxial tension or there are no cracks on the tensile surface of a flat specimen upon testing for a four-point bend.

In accordance with the Gazprom 2-4.1-223-2008 standard, the SSC and HIC resistance of the base metal and the welded joint of the pipes are classified into four groups (Table 1.6).

The API Spec 5L and ISO 3183-3 specifications provide the following requirements for HIC and SSC for the PSL-2 pipe steel:

- CLR ≤ 15%;
- CTR ≤ 5%;
- CSR ≤ 2%;
- four-point bending specimen should be tested at a stress of minimum 72% of the specified minimum yield strength of the pipe.

Table 1.4 Requirements for the results of tensile tests of pipes ordered for sour service according to API Spec 5L (API Spec 5L 2012) and ISO 3183-3 (ISO 3183-3 2012)

Pipe steel grade	Pipe body of SMLS and welded pipes					Weld seam of HFW and SAW pipes
	Yield strength $R_{t0.5}$ (MPa)		Tensile strength R_m (MPa)		Ratio $R_{t0.5}/R_m$	Tensile strength R_m (MPa)
	Min.	Max.	Min.	Max.	Min.	Min.
L245NS or BNS L245QS or BQS L245MS or BMS	245	450	415	655	0.93	415
L290NS or X42NS L290QS or X42QS L290MS or X42MS	290	495	415	655	0.93	415
L320NS or X46NS L320QS or X46QS L320MS or X46MS	320	525	435	655	0.93	435
L360NS or X52NS L360QS or X52QS L360MS or X52MS	360	530	460	760	0.93	460
L390QS or X56QS L390MS or X56MS	390	545	490	760	0.93	490
L415QS or X60QS L415MS or X60MS	415	565	520	760	0.93	520
L450QS or X65QS L450MS or X65MS	450	600	535	760	0.93	535
L485QS or X70QS L485MS or X70MS	485	635	570	760	0.93	570

The minimum elongation is calculated according to the minimum yield strength

Table 1.5 Requirements for the mechanical properties of pipes according to STO Gazprom 2-4.1-223-2008 (STO Gazprom 2008)

Grade	σ_B (N/mm^2)	$\sigma_{0.2}$ (N/mm^2)	δ_5 (%)	Impact toughness (J/cm^2)		DWTT^{-20} shear area (%)
				KCV^{-20}	KCU^{-60}	
Not less than						
K48	470	265	20	49	49	50
K50	485	343				
K52	510	353				
X42SS	414	290				
X46SS	434	317				
X52SS	455	359				

1. The maximum ultimate tensile strength should not exceed the nominal one by more than 118 N/mm^2

2. The $\sigma_{0.2}/\sigma_B$ ratio should not exceed 0.88 for pipes from controlled rolled plates and 0.85 in other cases

Table 1.6 Corrosion cracking resistance parameters of pipe metal according to STO Gazprom 2-4.1-223-2008 (STO Gazprom 2008)

Resistance group	Resistance parameters		
	SSC (method A)	HIC	
	σ_{thr} in fractions of $\sigma_{0.2\ min}$, not less	CLR (%), maximum	CTR (%), maximum
C-1	0.8	3	$\rightarrow 0$[a]
C-2	0.7	6	1
C-3	0.6	12	2
C-4	0.5	20	3

[a]Isolated small cracks are present in the same plane or in the planes unequally distant from the surface, are not connected by transverse crack with the formation of steps, and are distant from each other over a spacing of more than 0.5 mm

The HIC and SSC tests are performed in Solution A. The steels for submarine pipelines and, in some cases, high-strength pipes not intended for use in sour environments are tested for cracking resistance in less aggressive Solution B.

In some cases, no blisters are permitted on the surface of specimens after HIC test.

The general corrosion rate can be specified for hydrogen sulfide-resistant pipes. The tests are carried out in model media containing H_2S or CO_2. The corrosion rate should not exceed 0.40–0.50 mm/year.

The pipes from carbon steels and low-alloy steels are not resistant to massive hydrogen sulfide corrosion. Therefore, the protection against such kind of damage can be provided by using corrosion inhibitors, coatings, gas drying, and other methods, and monitoring of its development should be carried out throughout the entire service life in accordance with operation rules.

References

ANSI/NACE. (2011). Standard TM0284–2011. *Standard Test Method «Evaluation of Pipeline and Pressure Vessel Steels for Resistance to Hydrogen-Induced Cracking»*. Houston, p. 20

ANSI/NACE. (2016). *Standard TM0177–2016. Standard test method «Laboratory testing of metals for resistance to sulfide stress cracking and stress corrosion cracking in H_2S environments»*, Houston (p. 77).

API Specification 5L. (2012). *Specification for line pipe* (45th ed., p. 164). Washington.

Bosch, C., Haase, T., Liessem, A., Jansen, J.-P. (2008). Effect of NACE TM0284 test modifications on the HIC performance of large-diameter pipes. In: *International NACE Corrosion Conference and Expo, New Orlean* (Vol. 2, p. 6553), March 16–20, 2008.

EN 10208-2:2009. *Steel pipes for pipelines for combustible fluids—Technical delivery conditions—Part 2: Pipes of requirement class B* (p. 56)

Haase, T., Bosch, C., Maerten, B., Schroeder, J. (2012). Influence of plastic strain on the resistance of heavy-wall SAWL large-diameter pipes to hydrogen induced cracking and sulfide stress cracking. In: *International NACE Corrosion Conference and Expo, Salt Lake City* (Vol. 3, pp. 2151–2164), March 11–15, 2012.

ISO 15156-2:2009. (2009). *Petroleum and natural gas industries—Materials for use in H_2S-containing environments in oil and gas production—P. 2: Cracking-resistant carbon and low alloy steels, and the use of cast irons* (2-nd ed., p. 46).

LLC "VNIIGAZ". (2000). *Instructions for the selection and use of materials for the manufacture of pipes for pipelines operating in hydrogen sulfide-containing media, Moscow* (p. 34).

Petroleum and Natural Gas Industries. (2007). *International standard ISO 3183:2012 Steel pipe for pipeline transportation systems* (p. 49).

SNiP 2.05.06-85. (1985). *Trunk Pipelines, Gosstroy USSR, Moscow* (p. 85).

STO Gazprom. (2008). *2-4.1-223-2008 Technical requirements for electrically welded hydrogen sulfide-resistant pipes, Moscow* (p. 15).

Shabalov, I. P., Matrosov, Yu I, Kholodnyi, A. A., et al. (2017). *Steel for gas and oil pipelines resistant to fracture in hydrogen sulphide-containing media*. Moscow: Metallurgizdat.

Chapter 2
Factors Affecting the Cracking Resistance of Pipe Steels in H₂S-Containing Media

The negative effect of aggressive hydrogen sulfide-containing wet media in transported gas and oil causes disruption of continuity and failure of pipes by the mechanisms of hydrogen-induced cracking (HIC) and sulfide stress cracking (SSC) and is determined by a number of external and internal factors. The external factors include the corrosive environment characteristics such as the concentration of hydrogen sulfide, carbon dioxide, and other impurities, acidity (pH) in the aqueous medium, H_2S partial pressure, temperature, duration of the corrosive medium action on the material, and external stresses. The internal factors are related to the metallurgical properties of the pipe metal and include the chemical composition and microstructure of steel, the shape and distribution of non-metallic inclusions, strength properties, and internal stresses (Shabalov et al. 2017). In this chapter, we consider the methods to increase the resistance of pipe steels to cracking under the action of H_2S-containing medium by the minimization and elimination of metallurgical factors causing the nucleation and propagation of hydrogen-induced cracks.

2.1 Hydrogen Absorption

One of the methods to increase the cracking resistance of steel under the effect of sour media is the limitation of hydrogen absorption by creating a protective layer between the metal and aggressive solution. The effect of alloying elements and microstructure on the rate of hydrogen absorption by steel was considered in (USINOR ACIERS 1987; Shabalov et al. 2003; Yamada et al. 1983; Park et al. 2008).

Figure 2.1 shows the effect of alloying with Mo, Ni, and Cr on the hydrogen penetration rate for the steel containing 0.10%C-1.0%Mn-0.3%Cu upon holding in a Solution with pH = 5 (USINOR ACIERS 1987). The addition of 0.3% copper effectively decreases and virtually suppresses the hydrogen absorption. Additional alloying with molybdenum or nickel reduces the positive role of copper. The addition

© Springer Nature Switzerland AG 2019
I. Shabalov et al., *Pipeline Steels for Sour Service*, Topics in Mining, Metallurgy and Materials Engineering, https://doi.org/10.1007/978-3-030-00647-1_2

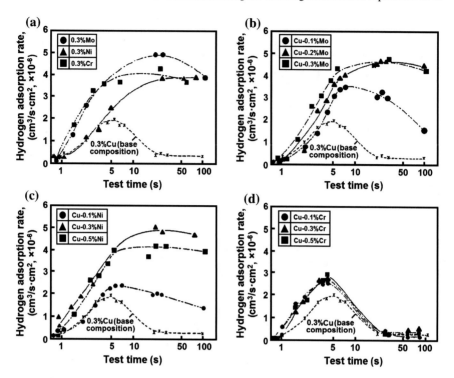

Fig. 2.1 Effect of additions of 0.3%Mo, Ni, or Cr (**a**) and concentrations of Mo (**b**), Ni (**c**), and Cr (**d**) on the rate of hydrogen absorption by the steel containing 0.10%C-1.0%Mn-0.3%Cu in Solution B (pH = 5) (USINOR ACIERS 1987)

of chromium into copper-containing steel does not virtually reduce the effect of copper.

Figure 2.2 shows the results demonstrating the existence of a correlation between the corrosion rate and hydrogenation (Shabalov et al. 2003). As the concentrations of copper, nickel, and chromium increase, the corrosion rate and the amount of hydrogen absorbed by the steel decrease.

The relationship between the corrosion rate and resistance to HIC of steel plates with different copper contents upon test in Solution B is shown in (Yamada et al. 1983) (Fig. 2.3). At a medium acidity of the environment (pH = 5), alloying with copper is an effective method for reducing the corrosion rate and increasing the HIC resistance of plates. As copper content is above 0.25%, the corrosion process in such environment virtually stops, and CLR tends to zero.

However, the protective surface layer formed due to alloying with copper is dissolved in media with lower pH 3-3.8 (Solution A), in which its effect on the hydrogen absorption by steel becomes negligible (Fig. 2.4) (Shabalov et al. 2017).

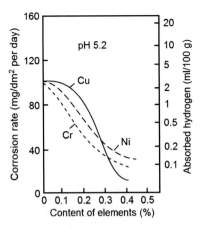

Fig. 2.2 Effect of the content of alloying elements (Cr, Ni, Cu) on the corrosion rate and the amount of absorbed hydrogen in Solution B (pH 5) (Shabalov et al. 2003)

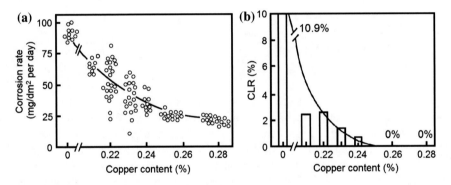

Fig. 2.3 Effect of copper content on the corrosion rate (**a**) and the crack length ratio CLR (**b**) upon testing X52 grade steel plates in Solution B (pH = 5) (Yamada et al. 1983)

An important factor determining the cracking susceptibility of steel in hydrogen sulfide-containing media is its microstructure affecting the solubility and concentration of hydrogen in steel. The entrapment of hydrogen and its diffusion depend on the type of the basic structure (steel matrix), high-carbon structure components, and non-metallic inclusions. Therefore, the effect of the microstructure character on the behavior of hydrogen upon its diffusion in steel is of great importance.

Depending on thermomechanical treatment regime, steels of the same chemical composition can differ in microstructure states and, correspondingly, in hydrogen permeability. Various microstructures were compared by the efficiency of hydrogen entrapment in (Park et al. 2008), and the hydrogen flux critical for HIC resistance was determined for the X65 grade low-alloy pipe steel of the chemical composition given in Table 2.1.

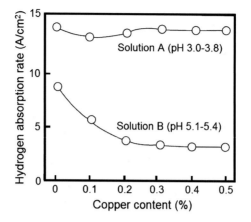

Fig. 2.4 Effect of copper content on the rate of hydrogen absorption by steel under the effect of media of different acidities (Shabalov et al. 2017)

Table 2.1 Chemical composition (wt%) of the X65 grade steel (Park et al. 2008)

C	Mn	Si	P	S	Nb + V + Ti	Cu + Ni	Cr	Mo	Al	Ca
0.05	1.25	0.20	0.005	0.002	~0.10	~0.40	~0.10	~0.10	~0.10	0.002

The slabs were rolled into plates 20 mm thick and cooled at a rate of 12 °C/s at various start and finish temperatures of accelerated cooling to form microstructures of different types (Fig. 2.5). The matrix of the test plates was ferrite, and the second phase was presented in the form of degenerate pearlite (DP), high-carbon upper bainite (UB), MA constituent (Fig. 2.6), and acicular bainitic ferrite (ABF).

The plates were made from the same slab. This provided a close level of the contamination by non-metallic inclusions (size, distribution, and fraction in area), which substantially affect the ability to hydrogen entrapment. The inclusions in the steel were identified as complex oxides of the Al–Ca–Si–O type with an average inclusion size of 1.5 μm, and the average fraction of the occupied area was 0.5%. Calcium treatment upon the steelmaking process provided the formation of oxide inclusions of small size and globular shape. No elongated manganese sulfides were observed.

The hydrogen permeability of the microstructures under examination was estimated from the hydrogen diffusion and entrapment parameters such as apparent permeability (D_{app}) and hydrogen solubility in steel (C_{app}). The D_{app} parameter characterizes the apparent diffusion permeability of the lattice with respect to dissolved and reversibly trapped hydrogen. The hydrogen solubility C_{app} corresponds to the hydrogen content in the lattice and in reversible traps. The lower the D_{app} and the higher the C_{app}, the more hydrogen is trapped by the steel. The test was performed in Solution A at an applied cathode current density of 500 μA cm^{-2}. The diffusion parameters and the results of the HIC test of the steel are given in Table 2.2.

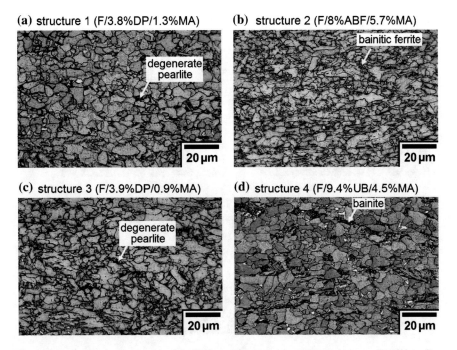

(a) structure 1 (F/3.8%DP/1.3%MA)   **(b)** structure 2 (F/8%ABF/5.7%MA)

(c) structure 3 (F/3.9%DP/0.9%MA)   **(d)** structure 4 (F/9.4%UB/4.5%MA)

Fig. 2.5 Types of microstructures of the X65 steel plates processed by various accelerated cooling regimes (Park et al. 2008)

Table 2.2 Results of the hydrogen permeability test and the HIC resistance of various microstructures of X65 grade steel plates (Park et al. 2008)

Microstructure version	Microstructure type	Hydrogen permeability parameters		HIC
		Apparent permeability D_{app} ($\times 10^{-10}$ m^2 s^{-1})	Hydrogen solubility C_{app} ($\times$ mol H m^{-3})	Crack area ratio (CAR) (%)
1	F/3.8% DP/1.3% MA	9.3	14.3	0
2	F/8.1% ABF/5.7% MA	4.1	20.9	0.05
3	F/3.9% DP/0.9% MA	9.4	13.8	0
4	F/9.4% HB/4.5% MA	4.4	27.1	0.57

The steel with the ABF structure (version 2) is characterized by lower D_{app} value and higher C_{app} compared to those of the steel with the F/DP microstructure (versions 1 and 3). The ABF in high-strength low-alloy steels is characterized by randomly oriented boundaries and high dislocation density, which could lead to low D_{app} and high C_{app}. This means that reversible hydrogen is more efficiently trapped by ABF than by the pearlite structure. It was concluded that the ABF and MA constituents are reversible traps (Park et al. 2008).

(a)   **(b)**

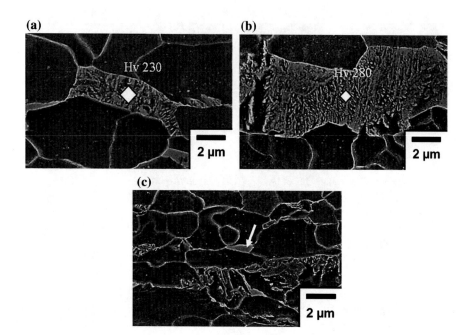

(c)

Fig. 2.6 High-carbon structure constituents in the microstructure of the plates: **a** degenerate pearlite; **b** high-carbon bainite; **c** MA constituent (Park et al. 2008)

The amount of hydrogen trapped depends on the type and fraction of the second phases acting as hydrogen trapping sites. For the pearlite structure, the pearlite boundaries and interfaces between ferrite and cementite in pearlite are the dominant trap sites of hydrogen. In bainite, the surfaces of a series of thin cementite plates act as inhibitors of hydrogen diffusion. The retained austenite in the MA constituent does not trap hydrogen, but the boundaries between the retained austenite and the martensitic layer can serve as such sites.

Locally agglomerated MA constituents easily embrittled by hydrogen are sites of the occurrence of hydrogen-induced cracks in steels with F/ABF or F/HB microstructure. As the microstructure changes from F/ABF to F/HB at comparable amounts of MA constituent, the sensitivity to HIC increases, which is explained by the high toughness of ABF, which prevents the propagation of the hydrogen-induced crack. The fraction of the second phase can be effectively controlled with the help of controlled rolling and post-deformation cooling.

2.2 Non-metallic Inclusions

The accumulation of atomic hydrogen, which saturates the steel, as well as its recombination into the molecular form occur at structure discontinuities such as interphase interfaces between non-metallic inclusions and steel matrix. This leads to the appearance of internal gas pressure and tensile stresses, which, after exceeding a certain critical level, initiate the crack nucleation. The stress level depends on the shape of the inclusions, their size, quantity, distribution, and distance from the plate/pipe surface. Due to the specific features of crystallization and solidification of a continuously cast slab, most non-metallic inclusions, harmful impurities (S, P), and alloying elements (Mn, Nb, C) are concentrated in the axial zone. Predominantly, this leads to hydrogen-induced cracking in the axial segregation region. Therefore, special attention upon the production of continuously cast slabs for plates intended for sour-gas-resistant pipes is paid to the measures aimed at increasing steel cleanliness from non-metallic inclusions and reducing centerline segregation heterogeneity (Shabalov et al. 2017).

The main non-metallic inclusions detrimentally affecting the cracking resistance of the plates of high-quality low-alloy microalloyed pipe steels in H_2S-containing media are as follows (Kholodnyy et al. 2017; Rykhlevskaya et al. 2006; Barykov 2016):

- inclusions based on manganese sulfide (MnS);
- inclusions of a complex composition based on niobium and titanium carbonitrides (Ti, Nb)(C, N);
- oxide–sulfide inclusions of complex composition.

The detrimental effect of the increased content of manganese sulfides on the HIC resistance of rolled products deserves the greatest attention. Manganese sulfides are formed in the slab upon solidification. A high concentration of MnS in the central zone of the slab is caused by an intense axial segregation of sulfur and manganese. Manganese sulfides have a high plasticity and are deformed during hot rolling, acquiring an elongated shape with sharp ends. High stresses can be induced at such inclusions under the pressure of molecular hydrogen and can cause crack nucleation.

The fractographic examination of the fracture surface of the base metal samples from the pipe (Rykhlevskaya et al. 2006) showed that the presence of a great number of MnS inclusions forces the metal to form hydrogen cracks (CLR = 5–21%). The absence or insignificant amount of MnS evidence a potentially high HIC resistance (CLR = 0–0.8%) of the metal.

Figure 2.7 shows elongated manganese sulfides on the polished section and on the fracture surface of hydrogen crack (Kholodnyy et al. 2017; Barykov 2016).

The effect of sulfur content on plate resistance to HIC is associated with the formation of inclusions based on manganese sulfides. Therefore, a high purity of steel relative to sulfur is a necessary condition for the HIC resistance of plates. Figure 2.8 shows that the sulfur content at pH 5 (test Solution B) should be maximum 0.002%, and at pH 3 (test Solution A) should be maximum 0.001% (CBMM/NPC 2001). Many researchers also note the need to provide a mass fraction of sulfur of maximum 0.001%.

(a)   **(b)**

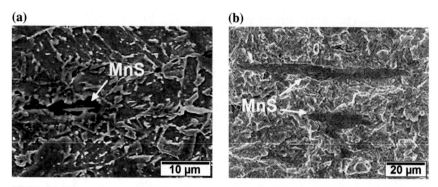

Fig.	Elements, wt. %						
	O	Mg	S	Nb	Mn	Fe	Cu
a	10.95	-	14.17	1.07	24.05	49.76	-
	10.55	-	13.45	0.95	22.76	52.29	-
b	-	9.93	41.26	-	43.97	4.84	-
	-	4.16	24.18	-	31.40	37.99	2.27

Fig. 2.7 Manganese sulfides (MnS) on the polished section (**a**) (Kholodnyy et al. 2017) and on the fracture surface of hydrogen crack (**b**) (Barykov 2016) in the axial zone of the plate and their chemical composition, SEM

Fig. 2.8 Effect of sulfur content on the crack sensitivity ratio (CSR) for testing X42 grade steel plates 5–10 mm thick in hydrogen sulfide-containing media differing in pH (CBMM/NPC 2001)

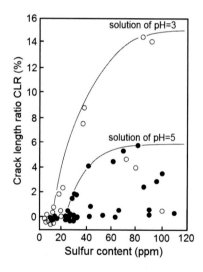

However, ensuring ultra-low sulfur content ($\leq 0.001\%$) in itself does not guarantee the prevention of manganese sulfide formation. It is necessary to effectively modify non-metallic inclusions in steel at the final stage of out-of-furnace treatment. The introduction of calcium into the steel allows one to minimize the amount of manganese sulfide particles and to provide their spherical shape (Fig. 2.9).

For the effective modifying treatment of non-metallic inclusions, the calcium content in the steel should exceed that corresponding to the Ca/S stoichiometric ratio.

(a) **(b)**

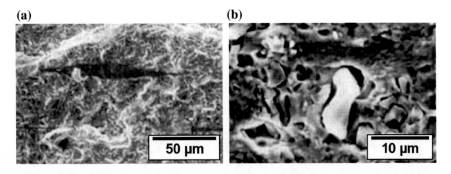

Fig. 2.9 Deformed manganese sulfides in the steel containing 0.0018%S without treatment with Ca (**a**) and the modified spherical inclusions in the calcium (0.0027%Ca)-treated steel containing 0.0003%S (**b**) at the fracture surface (Shabalov et al. 2017)

Fig. 2.10 Effect of the Ca/S ratio (at S ≤ 0.001%) on the frequency of CLR > 3% upon tests in a solution of pH = 3 (**a**) (CBMM/NPC 2001) and the hydrogen crack length upon tests in solutions of pH = 3.2 and 4.5 (**b**) (Shabalov et al. 2017)

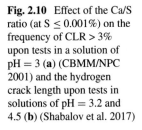

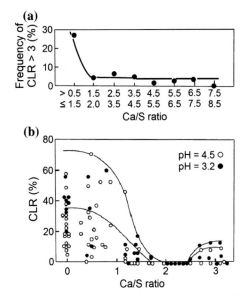

It was shown in (CBMM/NPC 2001) that, as the Ca/S ratio (at S ≤ 0.001%) exceeds 1.5, the frequency of CLR > 3% upon tests in a solution of pH = 3 substantially decreases (Fig. 2.10a). It was noted in Shabalov et al. (2017) that, at Ca/S = 2.0–2.5, CLR = 0 upon test in solutions of pH = 3.2 and 4.5 (Fig. 2.10b). Decrease or increase in the Ca/S ratio compared with its optimal value leads to hydrogen-induced cracking.

Along with calcium, rare earth additions such as cerium and lanthanum are used to modify manganese sulfides. The introduction of REM into the metal at the final stage of secondary treatment before casting in CCM substantially affects the composition of fine manganese sulfide inclusions concentrated in the mid-thickness zone of the plate and the efficiency of sulfur binding upon the metal solidification (Barykov 2016). Modification with cerium and lanthanum provides fine cleaning from sulfur

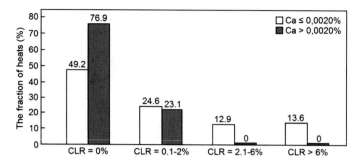

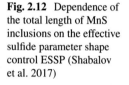

Fig. 2.11 Change in the CLR ratio as a function of the calcium content in metal before casting (REM consumption = const; S ≤ 0.001%) (Barykov 2016)

Fig. 2.12 Dependence of the total length of MnS inclusions on the effective sulfide parameter shape control ESSP (Shabalov et al. 2017)

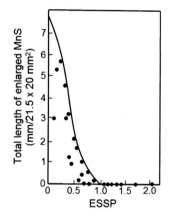

and oxygen and reduces the degree of contamination by non-metallic inclusions. This allows one to substantially increase the resistance of steel to sulfide stress cracking, hydrogen-induced cracking, and general corrosion in hydrogen sulfide-containing media (Denisova et al. 2013).

The results of HIC tests have been shown in (Barykov 2016) to depend on the calcium content in metal in the case where the sulfur content is ≤0.001%, and the consumption of the REM-containing modifier is same (Fig. 2.11). As the calcium content in the metal before casting was above 0.0020%, the fraction of heats with lower CLR ratio was higher at a Ca content of ≤0.0020%. However, if the calcium content is much higher than the required level, the excessive calcium can form oxide–sulfide inclusions of complex composition, and they will act as initiators of hydrogen cracking.

To prevent HIC, one should simultaneously control the calcium, oxygen, and sulfur contents using the effective sulfide shape control parameter ESSP = [Ca] × (1–124[O])/1.25[S]. If ESSP is >1.0, the shape of the MnS sulfides is completely globular, and there are no elongated MnS inclusions in the steel (Fig. 2.12) (Shabalov et al. 2017).

Fig. 2.13 (Ti, Nb)(C, N)
inclusion of in the
segregation band of rolled
plate

In modern high-quality pipe steels, the sulfur concentration ranges from <0.001 to 0.007%. This is primarily conditioned by the need to provide a high level of cold resistance of the steel.

Simultaneously with a decrease in the sulfur and manganese concentrations in steel, it is important to decrease the total degree of axial segregation of continuously cast slab for reducing the quantity of MnS.

Other non-metallic inclusions serving as initiators of cracking in steel in a hydrogen sulfide-containing medium are coarse inclusions (up to 5 μm) of complex composition based on (Ti, Nb)(C, N) carbonitrides of microalloying elements niobium and titanium (Shabalov et al. 2017; Barykov 2016). In the modern pipe steels microalloyed with Ti and Nb, such inclusions are formed upon solidification during continuous casting and can form coarse aggregates in the axial zone of the slab. Upon rolling, such hard inclusions create microdiscontinuities and increase the stress level in the surrounding metal. Figure 2.13 shows a coarse carbonitride particle in the axial zone of the plate.

Upon tests for cracking resistance in H_2S-containing media, hydrogen is retained at the (Ti, Nb)(C, N) particles and in the surrounding discontinuities and, recombinating into the molecular form, causes cracking. The aggregation of the (Ti, Nb)(C, N) particles on the surface of the polished section and at the fracture surface of hydrogen crack in the axial zone is shown in Fig. 2.14 (Kholodnyy et al. 2017; Barykov 2016). Figure 2.15 demonstrates a panoramic image of the hydrogen crack near the (Ti, Nb)(C, N) particles.

At increased C, Mn, and Nb contents, bainitic bands are formed in centerline segregation zone upon accelerated cooling. The presence of (Ti, Nb)(C, N) particles in such bands facilitates the propagation of hydrogen-induced cracks.

To reduce or eliminate the presence of coarse titanium and niobium carbonitrides in the axial zone, their mass fraction and the conditions of their introduction into liquid steel should be optimized. It has been established that niobium should be introduced into ladle furnace at a metal temperature of at least 1650 °C (Barykov 2016).

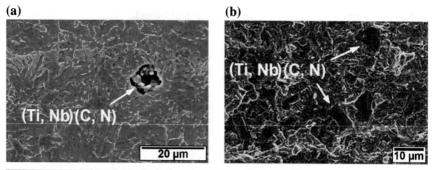

Fig.	Elements, wt. %								
	C	N	O	Ca	Ti	Nb	V	Mn	Fe
a	7.70	20.28	-	-	41.63	6.26	0.60	1.12	22.41
	3.05	12.74	8.52	0.28	26.97	4.77	0.36	0.63	42.68
	12.50	16.06	9.63	-	36.87	8.58	0.69	0.72	14.95
b	15.57	7.66	-	-	2.18	68.19	-	-	6.40
	19.79	6.55	-	-	2.03	64.65	-	-	6.98

Fig. 2.14 Aggregation of (Ti, Nb)(C, N) carbonitride inclusions on the surface of the polished section (**a**) (Kholodnyy et al. 2017) and on the fracture surface of the hydrogen crack (**b**) (Barykov 2016) in the centerline segregation zone of the plates and their chemical composition, SEM

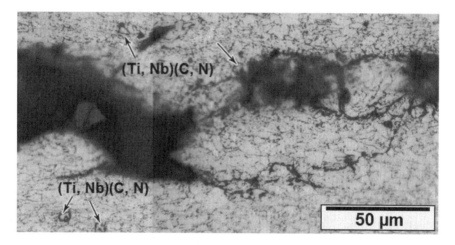

Fig. 2.15 Propagation of hydrogen cracks near the aggregates of the (Ti, Nb)(C, N) particles

Oxide–sulfide inclusions of a complex composition were found in the region of the concentration of hydrogen cracks. The oxide part in such inclusions is presented by aluminum–magnesium spinel, and the sulfides are presented by calcium, manganese, and, in some cases, REM (Fig. 2.16) (Barykov 2016).

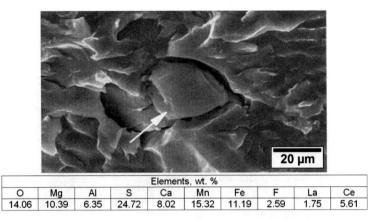

Elements, wt. %									
O	Mg	Al	S	Ca	Mn	Fe	F	La	Ce
14.06	10.39	6.35	24.72	8.02	15.32	11.19	2.59	1.75	5.61

Fig. 2.16 General view and composition of the MgAl$_2$O$_4$-based non-metallic inclusion, SEM (Barykov 2016)

Oxide inclusions are controlled upon the out-of-furnace treatment. To facilitate the separation of aluminum-containing particles in liquid steel, a soft purge by argon is usually used after verification that the upper layer of the slag is not damaged and no secondary oxidation occurs. The use of submerged entry nozzles between the tundish ladle and the intermediate ladle as well as between the intermediate ladle and the mold prevents a possible contact of the metal with air upon continuous casting. As the metal passes through the partitions in the intermediate ladle, the basic ladle slag facilitates further separation of the inclusions. All such measures provide oxygen content in the final product below 20 ppm, at an average value of about 11 ppm.

The distribution and density of non-metallic inclusions, especially oxides, over the slab thickness substantially depend on the CCM design (Lachmund and Bruckhaus 2006). Figure 2.17 shows the effect of the CCM type on the steel purity expressed as the density of non-metallic inclusions, and an example is given for the distribution of non-metallic inclusions over the thickness of slabs cast with CCMs of various designs.

It is seen that the smallest contamination by inclusions is characteristic of the slab cast with a continuous caster of vertical type. Reducing the length of the vertical section leads to an increase in the particle density. This is due to the fact that the inclusions in the vertical section float up and dissolve in the slag. In the curvilinear section, the inclusions floating up are captured by the solidifying part of the ingot, and the non-metallic inclusions are always concentrated near the upper part of the slab (Fig. 2.17b). At the same time, the amount of non-metallic inclusions in the steel cast with the CCM of vertical type is much smaller, and they are distributed more evenly over the slab volume. The use of short vertical sections about 2–3 m long can only slightly increase the removal of inclusions. High steel cleanliness from non-metallic inclusions is guaranteed after casting in CCM with a vertical section 12–15 m long.

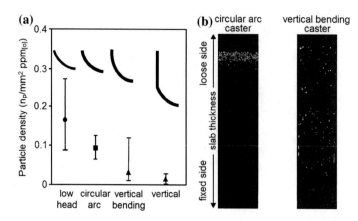

Fig. 2.17 Effect of CCM design on the steel purity from non-metallic inclusions: **a** particle density; **b** distribution of particles over the slab thickness (Lachmund and Bruckhaus 2006)

Fig. 2.18 Non-metallic inclusions and their composition in hydrogen sulfide-resistant pipe steels of X52 and X65 grades (Shabalov et al. 2017)

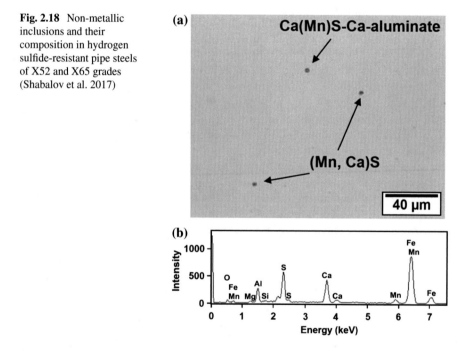

The results of the study of high-quality rolled coils and plates up to 20 mm thick from hydrogen sulfide-resistant pipe steels (CLR = 0%, CTR = 0%, CSR = 0%, SSC according to NACE TM 0177) of X52 and X65 grades are presented in (Shabalov et al. 2017). Figure 2.18 shows typical modified sulfide inclusions in experimental rolled steel.

The implementation of the manufacture of plates 20 mm thick from X52 to X65 grade steels is considered in (Kholodnyi et al. 2016). The plate metal was characterized by high cleanliness from non-metallic inclusions. The observed inclusions were globular, fine, and evenly distributed over the plate volume. It was noted that such inclusions do not substantially affect the cracking of steel in hydrogen sulfide-containing media. Numerous fine non-deformed inclusions do not lead to the formation of cracks upon HIC tests.

2.3 Microstructure of Steel

The modern level of metallurgical technology of steelmaking, out-of-furnace processing, and continuous casting can provide minimum contamination of the slab by non-metallic inclusions. Such cleanliness of the steel is an essential, but not final condition of the high HIC resistance of plates. The other important factor is the type of microstructure formed in the steel at the stage of thermomechanical processing. The effect of the microstructure state of steel on the tendency to hydrogen-induced cracking was studied in (Shabalov et al. 2017; Kholodnyy et al. 2017). The X60 grade plates 20 mm thick were produced from the slabs of the same steel heat containing 0.07%C, 0.20%Si, 0.95%Mn, 0.25%Cr, 0.25%Ni, 0.25%Cu, and Ti + Nb + V $\leq$ 0.12%. The concentration of harmful impurities S and P was as low as $\leq$0.001 and $\leq$0.012%, respectively. At the stages of melting and continuous casting, the measures were undertaken to reduce the contamination by non-metallic inclusions and to minimize the segregation heterogeneity of the slabs.

The effect of the microstructure of the base metal and centerline segregation zone on the HIC resistance was determined by the examination of the plates treated by various regimes of controlled rolling with accelerated cooling:

- regime 1: finish of deformation and start of accelerated cooling in the single-phase γ field with the finish of cooling at 550 °C;
- regime 2: finish of deformation and start of accelerated cooling in the two-phase $(\gamma + \alpha)$ field with the finish of cooling at a lower temperature of 415 °C.

The cooling rate of the plates was 25–30 °C/s.

The content of non-metallic inclusions in the steel of the test plates was very small. The degree of contamination by non-metallic inclusions for the plates made from slabs of the same heat was of the same magnitude. Figures 2.19 and 2.20 show the distribution, type, and composition of the inclusions. Single isolated globular inclusions evenly distributed over the volume of plates are seen on the polished sections.

High contents of O, Mg, Al, S, Ca, and Mn in the chemical composition of the inclusions (see Fig. 2.20) allow one to identify them as complex inclusions of manganese oxysulfide and alumomagnesium spinel. Neither deformed manganese sulfides nor coarse (Ti, Nb)(C, N) carbonitride particles have been detected. A high cleanliness of the steel from non-metallic inclusions allowed one to exclude them

Fig. 2.19 Distribution of non-metallic inclusions on the surface of the unetched polished section of the sample taken from steel plate manufactured by regime 2

(a)   **(b)**

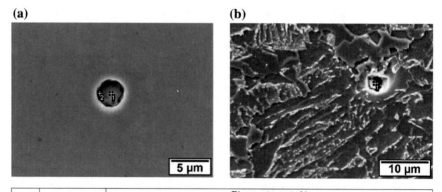

Fig.	Spectrum	Elements, wt. %							
		O	Mg	Al	S	Ca	Ti	Mn	Fe
a	1	41.52	4.61	30.24	2.75	2.92	0.52	1.33	16.11
	2	13.34	2.16	12.58	9.22	8.60	-	3.17	50.93
b	1	6.03	3.82	5.12	7.45	7.44	2.66	0.99	66.49

Fig. 2.20 Complex inclusions of manganese oxysulfide and alumomagnesium spinel and their chemical compositions in the plate manufactured by regime 2, SEM: **a** unetched polished section; **b** etched polished section, SEM

from the factors significantly affecting the cracking resistance of the experimental plates in hydrogen sulfide-containing media.

Figure 2.21 shows the microstructure of the base metal of the plates.

After processing by both regimes of controlled rolling with accelerated cooling, the base metal of the plates was free from structure non-uniformity in the form of banding, which can be the site of nucleation and propagation of hydrogen-induced cracks.

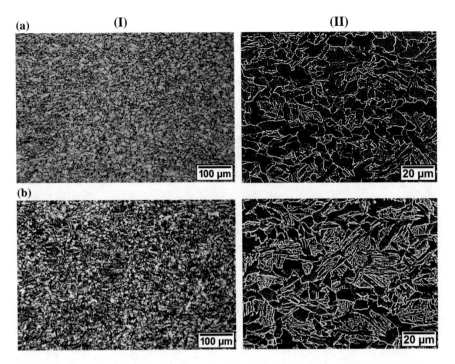

Fig. 2.21 Microstructure of the plates base metal, I—OM; II—SEM: **a** regime 1; **b** regime 2

Figure 2.22 shows the microstructure of the centerline segregation zone of the test plates. The plates manufactured by regime 1 are characterized by a high degree of homogeneity over thickness. Hardly discernible segregation bands are visible in the axial zone (Fig. 2.22a). The plate processed by regime 2 exhibits a substantial structural heterogeneity expressed by coarse segregation bands in the axial zone compared to the base metal (Fig. 2.22b).

The microhardness of the base metal of both plates was the same, 178–184 $HV_{0.2}$ (Table 2.3). The microhardness of the centerline segregation zone was higher than that of the base metal. The microhardness of the plate treated by regime 1 was 223 $HV_{0.2}$, while that of the plate treated by regime 2 was much higher, 305 $HV_{0.2}$. The difference between the microhardnesses of the centerline segregation zone and base metal ($\Delta HV_{0.2}$) for the plates treated by regimes 1 and 2 was 45 $HV_{0.2}$ and 121 $HV_{0.2}$, respectively. The segregation structural heterogeneity coefficient expressed as the ratio between the microhardnesses of the axial zone and the base metal, $K(HV_{0.2})$, was 1.25 and 1.66, respectively.

In view of the difference in the microstructure and hardness of the base metal and the axial zone, such regions of the plate metal were compared for the susceptibility to HIC. Figure 2.23 shows the specimen cutting scheme. Standard full-thickness specimens were cut from the plate 20 mm thick for the HIC test, and the specimens

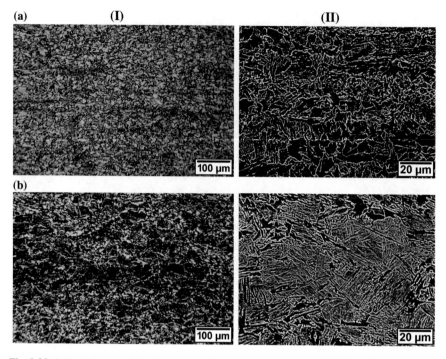

Fig. 2.22 Microstructure of the plates centerline segregation zone, I—OM; II—SEM: **a** regime 1; **b** regime 2

Table 2.3 Microhardness of the base metal and centerline segregation zone and the $\Delta HV_{0.2}$ and $K(HV_{0.2})$ parameters of plates

CR + AC regime	Microhardness $HV_{0.2}$		$\Delta HV_{0.2}$	$K(HV_{0.2})$
	Base metal	Centerline zone		
1	178	223	45	1.25
2	184	305	121	1.66

8 mm thick were cut from the segregation zone at the middle region over the plate thickness and from the base metal free from segregation zone.

The results of plate tests for HIC resistance are presented in Table 2.4. In the case of testing specimens taken from the plate treated by regime 1, no hydrogen cracking is observed both upon tests of standard full-thickness HIC specimens and upon tests of specimens of reduced thickness with and without segregation zone. On the contrary, the tests of full-thickness specimens and of the specimens taken from the axial zone of the plate treated by regime 2 exhibited hydrogen-induced cracking. Upon testing, the specimen cut from the base metal without segregation zone, no HIC was observed. The hydrogen-induced cracks were observed in the axial zone of the plate, and they propagated along the segregation bands of increased hardness (Fig. 2.24).

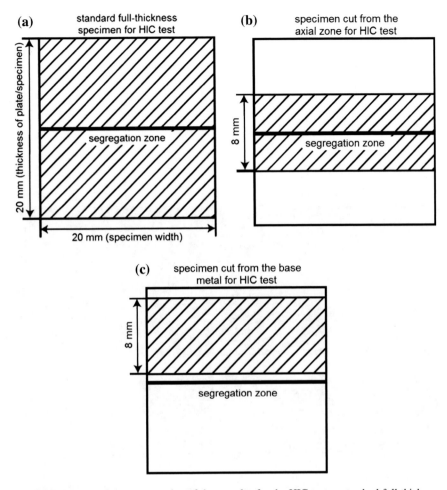

Fig. 2.23 Scheme of the cross section of the samples for the HIC test: **a** standard full-thickness specimen; **b** specimen cut at ½ of the thickness of plate with the segregation zone; **c** specimen cut at ¼ of the thickness of plate free from segregation zone

The fractographic study showed a brittle character of the surface of hydrogen-induced crack (Fig. 2.25). The fracture surface exhibited rare non-metallic inclusions in the form of fine modified globular complex particles of manganese oxysulfide and aluminum–magnesium spinel (see Fig. 2.25a). No association of such inclusions with the nucleation and propagation of hydrogen-induced cracks was detected.

According to the results of the study, it was concluded that the predominant sites of the nucleation and propagation of hydrogen cracks in the plates of high cleanliness from non-metallic inclusions are extended segregation bands of increased hardness in the zone of central structural heterogeneity of the plates. Microstructure of the

Table 2.4 Results of the plate tests for HIC resistance

CR + AC regime	HIC parameter (%)	Full-thickness specimen	Specimen cut from the segregation zone	Specimen cut from the base metal
1	CLR	0	0	0
	CTR	0	0	0
	CSR	0	0	0
2	CLR	19.6	18.1	0
	CTR	0.97	5.0	0
	CSR	0.183	0.90	0

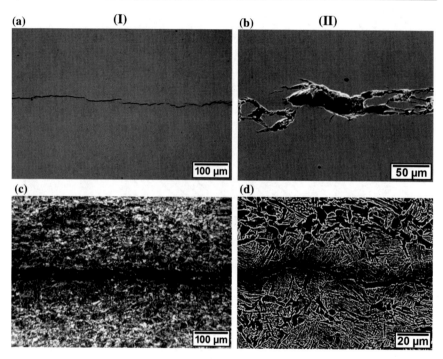

Fig. 2.24 Hydrogen-induced crack in the centerline segregation zone of the plate processed by regime 2, I—OM; II—SEM: **a, b** unetched polished section; **c, d** polished section after etching

base metal free from structural heterogeneity in the form of banding does not show a tendency to hydrogen cracking.

The tendency to HIC was estimated for various plate microstructures. Figures 2.26, 2.27, 2.28, 2.29, and 2.30 show that, in the plates produced by different thermomechanical processing regimes, the hydrogen-induced cracks propagate in the zone of central structural heterogeneity along coarse extended segregation bands with regions of high-carbon structures such as lamellar pearlite, degenerate pearlite, high-carbon

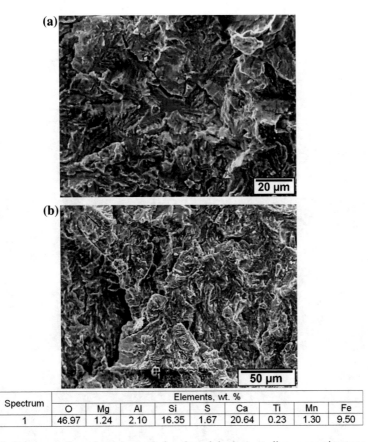

Spectrum	Elements, wt. %								
	O	Mg	Al	Si	S	Ca	Ti	Mn	Fe
1	46.97	1.24	2.10	16.35	1.67	20.64	0.23	1.30	9.50

Fig. 2.25 Fracture surface of a hydrogen-induced crack in the centerline segregation zone of the plate processed by regime 2, SEM

upper bainite, twinned high-carbon martensite with retained austenite, and acicular bainitic ferrite with interlayers of retained austenite along the lath boundaries.

As the plate metal is manufactured with the completion of deformation in a finishing stand in the two-phase $(\gamma + \alpha)$ region, the formation of banded structure can occur not only in the central segregation zone, but also in the base metal. Such elongated bands in the base metal can be also the sites of the nucleation and propagation of hydrogen cracks. Figure 2.31 shows hydrogen cracks propagating along the pearlite bands in the base metal of the plates produced from the steel, which is non-resistant to hydrogen sulfide cracking.

To prevent HIC in the base metal, it is necessary to prevent the banded microstructure formation in the plate. This is promoted by a reduction in the carbon content, an increase in the end temperature of finishing rolling, and the use of accelerated cooling.

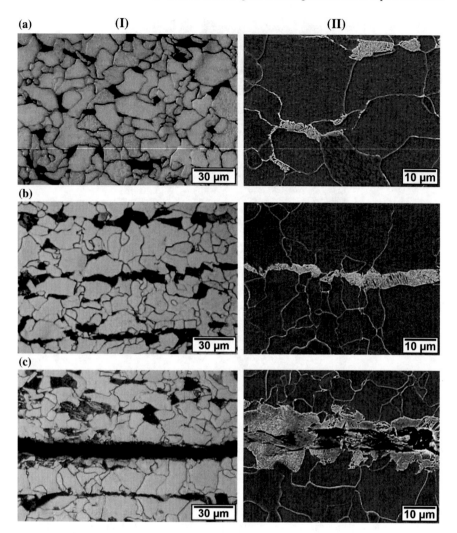

Fig. 2.26 Plate microstructure consisting of ferrite matrix with lamellar pearlite regions, I—OM; II—SEM: **a** uniform distribution of lamellar pearlite regions (base metal); **b** elongated segregation bands consisting of lamellar pearlite regions (axial zone); **c** hydrogen-induced cracks propagating along the elongated segregation bands consisting of lamellar pearlite regions (axial zone)

2.4   Strength Properties

Segregation bands in the axial zone of plates are characterized by an increased content of strongly segregating elements and, accordingly, a higher hardness relative to that of the base metal. To increase the HIC resistance of plates, it is necessary to reduce the central segregation of slab. The effect of the initial center segregation of the

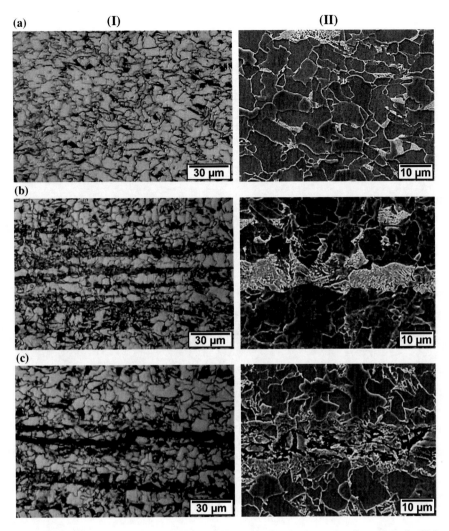

Fig. 2.27 Plate microstructure consisting of ferrite matrix with degenerate pearlite regions, I—OM; II—SEM: **a** uniform distribution of degenerate pearlite regions (base metal); **b** elongated segregation bands consisting of degenerate pearlite regions (axial zone); **c** hydrogen-induced cracks propagating along the elongated segregation bands consisting of degenerate pearlite regions (axial zone)

slabs from the steels containing 0.04–0.08%C and 0.65–1.35%Mn on the central segregation structural heterogeneity and hydrogen cracking resistance of the plates was established in (Shabalov et al. 2017). The plates 18–25 mm thick after controlled rolling were subjected to accelerated cooling by the same regimes: $T_{sc} = Ar_3 +$ (0–30) °C, $T_{fc} = 490$–550 °C, $V_c = 22$–30 °C/s. Due to reduced concentration of carbon and manganese, the central segregation heterogeneity of a continuously cast blank was decreased from a number of 2.5 to a number of 1.5 estimated by the

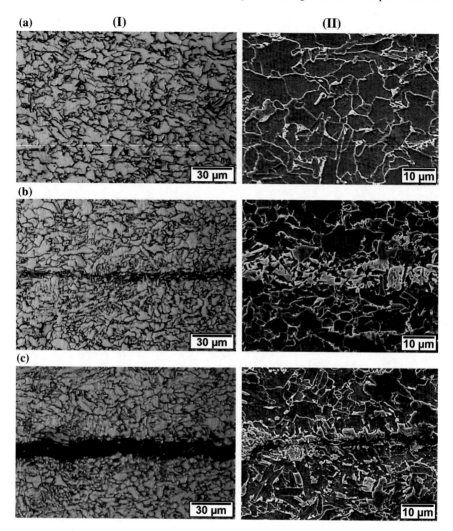

Fig. 2.28 Plate microstructure consisting of ferrite matrix with high-carbon upper bainite regions, I—OM; II—SEM: **a** uniform distribution of high-carbon upper bainite regions (basic metal); **b** elongated segregation bands consisting of high-carbon upper bainite regions (axial zone); **c** hydrogen-induced cracks propagating along the elongated segregation bands consisting of high-carbon upper bainite regions (axial zone)

Mannesmann method. As a result, the microhardness of the axial zone of the plates decreased by 135 $HV_{0.2}$, from 325 to 190 $HV_{0.2}$, while the microhardness of the base metal decreased only by 25 $HV_{0.2}$ (from 200 to 175 $HV_{0.2}$).

The plates were tested for HIC resistance. All detected hydrogen-induced cracks propagate in the segregation zone of the plates. Figure 2.32 shows the effect of the microhardness of the plate axial zone on the HIC parameters. It is seen from the

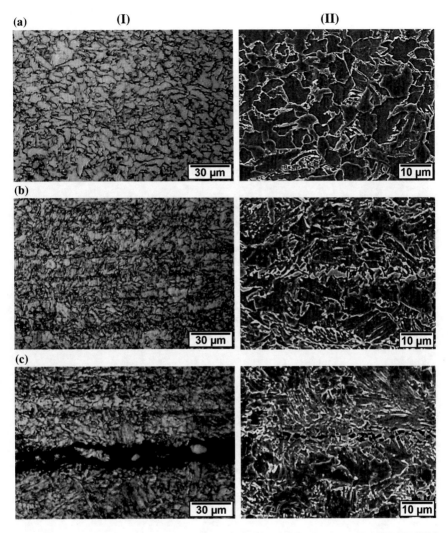

Fig. 2.29 Plate microstructure consisting of ferrite matrix with MA constituent, I—OM; II—SEM: **a** uniform distribution of the MA constituent (base metal); **b** extended segregation bands consisting of the MA constituent (axial zone); **c** hydrogen-induced cracks propagating along extended segregation bands consisting of the MA constituent (axial zone)

graphs that, as the microhardness of the axial zone decreases, the HIC resistance of the plate increases. At the same time, no HIC occurs as the microhardness of the axial zone is ≤ 225 $HV_{0.2}$.

It has been established that, as the plates are processed by the experimental CR + AC regimes, an axial zone microhardness of ≤ 225 $HV_{0.2}$ can be provided in the plates rolled from slabs with a central segregation heterogeneity of ≤ 2 in number. It was noted (CBMM/NPC 2001) that hydrogen cracking can be avoided if the

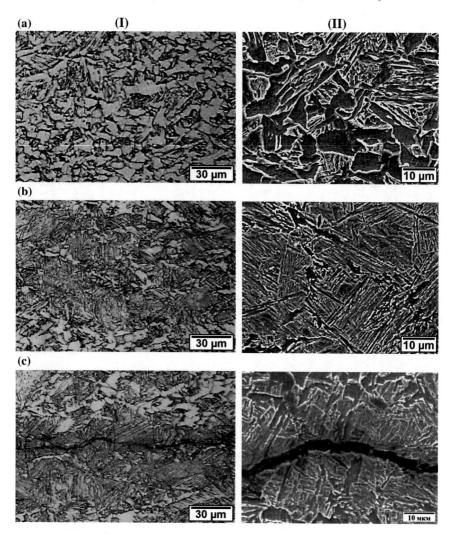

Fig. 2.30 Plate microstructure consisting of ferrite matrix and acicular bainitic ferrite (ABF) regions with interlayers of retained austenite (A_{ret}) at the lath boundaries, I—OM; II—SEM: **a** uniform distribution of ABF regions with A_{ret} interlayers (base metal); **b** extended segregation bands consisting of ABF regions with A_{ret} interlayers (axial zone); **c** hydrogen-induced cracks propagating along extended segregation bands consisting of ABF regions with A_{ret} interlayers (axial zone)

microhardness of the segregation band is below 330 HV, while in (Shabalov et al. 2017) the critical microhardness, at which the hydrogen crack is nucleated, was noted to be 250 HV both in the central zone and in the base metal. In the plates manufactured using controlled rolling technology with cooling in air, the microstructure of the axial zone with pearlite segregation bands of 180 $HV_{0.2}$ in microhardness exhibits a low HIC resistance.

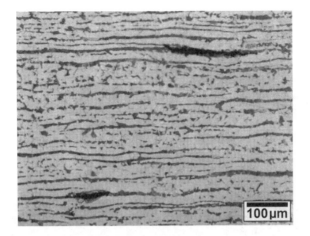

Fig. 2.31 Microstructure of the base metal of the plate with hydrogen cracks propagating along the pearlite bands in the base metal (Shabalov et al. 2017)

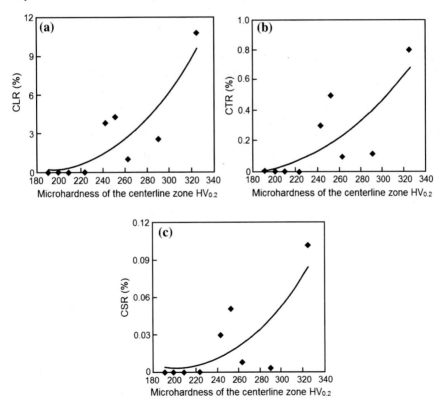

Fig. 2.32 Effect of the microhardness of the plate centerline zone on HIC parameters

The uniaxial tensile tests of cylindrical specimens (method A) or the four-point bending test of planar specimens are used for the evaluation of the SSC resistance of low-alloy pipe steels. The uniaxial tensile test specifying the holding of a preloaded specimen in a hydrogen sulfide-containing solution for 720 h is more severe. The SSC susceptibility of steel is determined by fracture. The threshold stress σ_{thr}, at which no failure by the SSC mechanism occurs, is determined by testing a series of specimens at different load levels.

The load acting on the specimen is determined by the following equation:

$$p = k \cdot \sigma_{T\,spec} \cdot S,$$

where

P is the load acting on the specimen (N);
k is the load factor;
$\sigma_{T\,spec}$ is the specified minimum yield strength (N/mm^2);
S is the minimum cross-sectional area of the gage section of the specimen (mm^2).

As follows from the equation, the load applied to the specimen depends on the load factor determined by the requirements imposed on the SSC resistance of steel and the specified minimum yield strength of the steel.

The results of testing specimens from industrial plates 20–23.8 mm thick from the X42 to X70 grade low-carbon low-alloy pipe steels (Table 2.5) have been analyzed to determine the effect of the load factor and strength properties of the steel on the SSC resistance under uniaxial tension of cylindrical specimens.

A low content of segregating elements (0.04–0.07%C, 0.90–1.35%Mn, 0.001%S, and 0.009–0.012%P) in the steels was necessary to ensure a high cracking resistance in H$_2$S-containing media. Alloying and microalloying were performed by additions of ≤0.75% (Cr + Ni + Cu) and ≤0.130% (Ti + Nb + V), respectively. The steels were characterized by a high purity due to a low content of harmful impurities and non-metallic inclusion.

Thermomechanical treatment of the slabs was performed in a plate mill using the technology of controlled rolling followed by accelerated cooling. The slabs were heated to temperatures of 1160–1180 °C for 4–6 h. Rough rolling was performed at temperatures from 1050–1100 °C to 980–1020 °C to a billet 100 mm thick. At the finishing stage, the start rolling temperature ranged between 840 and 920 °C and the finish rolling temperature was in a range of 775–865 °C. The post-deformation controlled cooling at a rate of 12–30 °C/s was started at 750–835 °C and finished at 430–625 °C.

The tensile properties of the metal were tested in accordance with ASTM A370 on full-thickness specimens taken in the transverse direction. The SSC resistance was tested in Solution A at load factors $k = 0.6, 0.7, 0.72, 0.8, 0.9$, and 1.0 from $\sigma_{T\,spec} = 380$ and 425 N/mm^2.

To determine the effect of the load factor k on the SSC resistance, the results of testing the samples from the steel 1 plates 20 mm thick from the slabs of the same

Table 2.5 Chemical composition of the steels

Steel	C	Mn	Si	S	P	Cr + Ni + Cu	Mo	Ti + Nb + V	Al	N	C_{eq}	P_{cm}
steel 1	0.07	0.90	0.17–0.22	0.001	0.009–0.012	≤ 0.75	0.01	≤ 0.130	0.031	0.009	0.33	0.16
steel 2	0.04–0.07	1.15–1.35					0.04–0.07		0.030–0.035	0.006–0.009	0.36–0.41	0.14–0.18

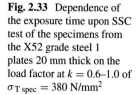

Fig. 2.33 Dependence of the exposure time upon SSC test of the specimens from the X52 grade steel 1 plates 20 mm thick on the load factor at $k = 0.6$–1.0 of $\sigma_{T\,spec} = 380$ N/mm^2

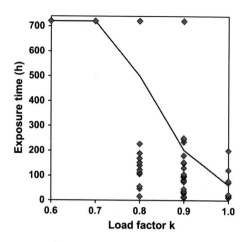

heat were analyzed. The regimes of controlled rolling ($T_{sr} = 910$–920 °C, $T_{fr} = 850$–865 °C) and accelerated cooling ($T_{sc} = 820$–835 °C, $T_{fc} = 485$–555 °C, $V_c = 23$–27 °C/s) were maintained in narrow ranges, which provided the same microstructure and properties of the plates under study. The actual tensile mechanical properties of the steel corresponded to the X52 grade and were within the following limits: $\sigma_{0.5} = 425$–455 N/mm^2, $\sigma_B = 515$–560 N/mm^2, $\delta_{2''} = 44$–54%, and $\sigma_{0.5}/\sigma_B = 0.80$–0.86. The plates had fine ferritic–bainitic microstructure consisting of quasipolygonal ferrite, a small fraction of uniformly distributed regions of high-carbon bainite, and cementite at ferrite grain boundaries.

The SSC resistance test was carried out at load factors $k = 0.6, 0.7, 0.8, 0.9$ (30 specimens each) and 1.0 (9 specimens) from $\sigma_{T\,spec} = 380$ N/mm^2 (Fig. 2.33).

All specimens tested at $k = 0.6$ and 0.7 passed the test and did not fail for 720 h. An increase in the load factor led to a decrease in the SSC resistance of the plates. This was expressed by a decrease in the average exposure time of the specimens. At the same time, the fraction of the failed specimens increased: Their numbers at $k = 0.8, 0.9$, and 1.0 were 11 (37%), 25 (83%), and 9 (100%), respectively. The exposure time of the failed specimens did not exceed 250 h and on average decreased with increasing load factor.

The effect of strength properties on the SSC resistance of steel was studied by testing the specimens from the steel 2 plates 20.6–23.8 mm thick. To obtain a wide range of properties, the plates were treated by various regimes of controlled rolling ($T_{sr} = 840$–905 °C, $T_{fr} = 775$–850 °C) and accelerated cooling ($T_{sc} = 750$–815 °C, $T_{fc} = 430$–625 °C, $V_c = 12$–30 °C/s). The tensile mechanical properties corresponded to the X42–X70 grades and were within the following limits: $\sigma_{0.5} = 445$–535 N/mm^2, $\sigma_B = 505$–635 N/mm^2, $\delta_{2''} = 40$–55%, $\sigma_{0.5}/\sigma_B = 0.81$–0.90. At high finish temperatures of accelerated cooling and low cooling rates, the plates exhibited the lowest strength properties and a ferritic–pearlitic microstructure. A decrease in the finish temperature of the accelerated cooling and increase in the cooling rate led to an increase in strength and the formation of a finer ferritic–bainitic microstructure.

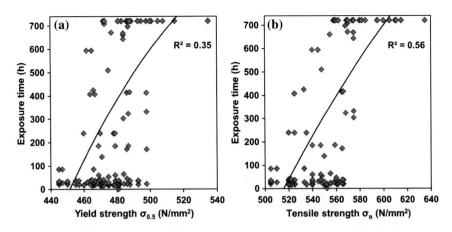

Fig. 2.34 Effect of the yield strength (**a**) and the tensile strength (**b**) of the specimens from the steel 2 plates 20.6–23.8 mm thick on the exposure time upon SSC tests at $k = 0.72$ of $\sigma_{T\,spec} = 425$ N/mm^2

Figure 2.34 shows the effect of the yield strength and tensile strength of the test plates on the exposure time of the specimens tested for SSC resistance at a load of $k = 0.72$ of $\sigma_{T\,spec} = 425$ N/mm^2. With increasing yield strength and tensile strength, the exposure time of the specimens increased. All specimens of the steel with a yield strength of >498 N/mm^2 and a tensile strength of >575 N/mm^2 did not fail for 720 h. At a yield strength of <472 N/mm^2 and an ultimate tensile strength of <558 N/mm^2, all specimens failed within less than 720 h. A strong dependence of the exposure time of the specimens on the tensile strength is revealed. A weaker dependence of the exposure time of the specimens on the yield strength is due to the fact that this parameter can decrease with the disappearance of the yield plateau in the stress–strain diagram as a result of changes in the microstructural state of the rolled products treated by different thermomechanical processing regimes. The tensile strength in this case gradually increases.

The effect of the strength properties of steel on the SSC resistance was studied at various load factors $k = 0.6–1.0$ of $\sigma_{T\,spec} = 425$ N/mm^2 (Fig. 2.35).

The strength characteristics of the test plate No. 1 ($\sigma_{0.5} = 460$ N/mm^2, $\sigma_B = 520$ N/mm^2) were lower than those of plate No. 2 ($\sigma_{0.5} = 505$ N/mm^2, $\sigma_B = 610$ N/mm^2). Upon testing specimens from both test plates at a load factor of 0.6, no failure was observed for 720 h. The increase in strength positively affected the SSC resistance of steel at higher load factors. All specimens from plate No. 1 failed upon testing at $k = 0.72, 0.8, 0.9$, and 1.0, and the specimens from the stronger plate No. 2 failed only at $k = 0.9$ and 1.0.

To estimate the effect of the safety factor of steel on the SSC resistance with allowance for the actual test load and the actual strength properties, the following parameters were used:

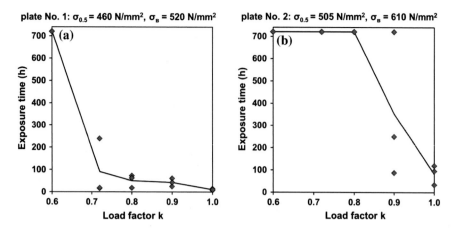

plate No. 1: $\sigma_{0.5}$ = 460 N/mm², σ_{B} = 520 N/mm²   **plate No. 2: $\sigma_{0.5}$ = 505 N/mm², σ_{B} = 610 N/mm²**

Fig. 2.35 Dependence of the exposure time of the specimens from the steel 2 plates 22 mm thick on the load factor of the SSC tests at $k = 0.6$–1.0 of $\sigma_{T\,spec} = 425$ N/mm²

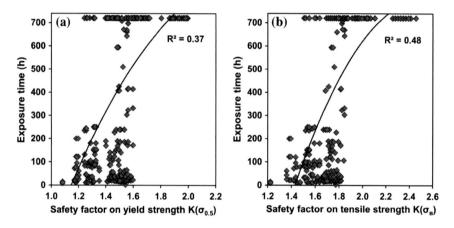

Fig. 2.36 Effect of the safety factors on yield strength $K(\sigma_{0.5})$ (**a**) and on tensile strength $K(\sigma_B)$ (**b**) on the exposure time of the specimens tested for SSC at $k = 0.6$–1.0 of $\sigma_{T\,spec} = 380$ and 425 N/mm² for the X42–X70 grade plates 20–23.8 mm thick

- safety factor on yield strength: $K(\sigma_{0.5}) = \sigma_{0.5}/k \cdot \sigma_{T\,spec}$;
- safety factor on tensile strength: $K(\sigma_B) = \sigma_B/k \cdot \sigma_{T\,spec}$.

Figure 2.36 shows the effect of the $K(\sigma_{0.5})$ and $K(\sigma_B)$ factors on the exposure time of the specimens tested for SSC at $k = 0.6$–1.0 of $\sigma_{T\,spec} = 380$ and 425 N/mm² for the X42–X70 grade plates 20–23.8 mm thick.

It is seen that, if the yield strength and the tensile strength are higher than the test load $k \cdot \sigma_{T\,spec}$ by a factor of >1.60 and >1.85, respectively, the SSC resistance of the steel is high, and all specimens do not fail for 720 h. At $K(\sigma_{0.5}) < 1.24$ and $K(\sigma_B) < 1.51$, all specimens failed within ≤200 h. At $K(\sigma_{0.5}) = 1.24$–1.60 and

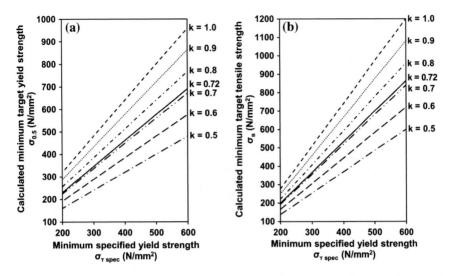

Fig. 2.37 Calculated minimum target yield strength (**a**) and tensile strength (**b**) providing SSC resistance of steel (exposure time $\geq$720 h) at load factors $k = 0.5$–1.0 of $\sigma_{T\,spec} = 200$–600 N/mm^2

$K(\sigma_B) = 1.51$–1.85, there are specimens failed and non-failed for 720 h. It follows that the high SSC resistance of the test steels is ensured at an actual yield strength $\sigma_{0.5} > 1.6 \cdot k \cdot \sigma_{T\,spec}$ and a tensile strength $\sigma_B > 1.85 \cdot k \cdot \sigma_{T\,spec}$.

On the basis of the revealed regularities, the minimum target yield strength and ultimate tensile strength values necessary to ensure a high SSC resistance upon test of the B-X80 grade steels at $k = 0.5$–1.0 of $\sigma_{T\,spec} = 200$–600 N/mm^2 have been calculated (Fig. 2.37). The established regularities can be used for the development of the technology for the production of plates and pipes of various strength grades and SSC resistance groups. When using the given target strength properties, the tensile strength is more indicative, since the yield strength can vary depending on the shape of the stress–strain diagram both upon thermomechanical processing and upon pipe manufacture.

The exposure time of the failed specimens was analyzed (Fig. 2.38). Half of all failed specimens exhibited an exposure time of $\leq$50 h. Of all specimens, 68% failed within 100 h, 76% failed within 150 h, 84% failed within 200 h, and 91% failed within 250 h. The remaining 9% of the specimens failed within 251–719 h. Of the specimens failed within $\leq$250 h, an average exposure time was 73 h. This shows that the most probable time to failure of the specimens is the initial test period. This may be caused by the fact that, just in this period, the saturation of the metal with hydrogen from the H$_2$S-containing solution is most intense.

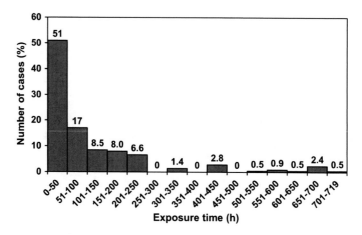

Fig. 2.38 Frequency distribution diagram of the exposure time of failed specimens tested for SSC at $k = 0.6$–1.0 of $\sigma_{T\,spec} = 380$ and 425 N/mm^2 for the X42–X70 grade steel plates 20–23.8 mm thick

References

Barykov A. B. (Ed.). (2016). *Development of steel manufacturing technology for steel rolled product and pipe in the Vyksa production area: Coll. Works. Metallurgizdat, Moscow.*

Denisova, T. V., Vyboishchik, M. A., Tetyueva, T. V., & Ioffe, A. V. (2013). Changes in the structure and properties of low-carbon low-alloy pipe steels upon inoculation with REM. *Metal Science and Heat Treatment, 54*(10), 530–534.

Kholodnyi, A. A., Sosin, S. V., Matrosov, Yu I, & Karmazin, A. V. (2016). Development of the production technology at the ISW "Azovstal" of sheets for hydrogen sulfide-resistant pipes of large diameter of the X52-X65 strength classes. *Problems of Ferrous Metallurgy and Materials Science, 4,* 26–34.

Kholodnyy, A. A., Matrosov, Y., Shabalov, I. P., et al. (2017). Factors influencing on resistance of pipe steels to cracking in H_2S-containing media. *Problems of Ferrous Metallurgy and Materials Science, 4,* 70–81.

Lachmund, H., Bruckhaus, R. (2006). Steelmaking process—basic requirement for sophisticated linepipe application. In: *Proceedings of the International Seminar on "Pipe Seminar Modern Steels for Gas and Oil Transmission Pipelines, Problems and Prospects", Moscow* (pp. 155–164). March 15–16, 2006.

Niobium Information. No. 18/01. (2001). *Sour gas resistant pipe steel, CBMM/NPC, Düsseldorf (Germany)*

Park, G. T., Koh, S. U., et al. (2008). Effect of microstructure on the hydrogen trapping efficiency and hydrogen induced cracking of linepipe steel. *Corrosion Science Journal, 50,* 1865–1871.

Rykhlevskaya, M. S., Platonov, S Yu., & Maslyanitsyn, V. A. (2006). Effect of special features of distribution of Non-metallic inclusions in the metal of electric-welded oil conducting tubes on resistance to hydrogen-stress cracking. *Metal Science and Heat Treatment, 48*(5), 448–451.

Shabalov, I. P., Matrosov, Yu I, Kholodnyi, A. A., et al. (2017). *Steel for gas and oil pipelines resistant to fracture in hydrogen sulphide-containing media.* Moscow: Metallurgizdat.

Shabalov, I. P., Morozov, Yu D., & Efron, L. I. (2003). *Steels for pipes and building structures with improved operating properties.* Moscow: Metallurgizdat.

USINOR ACIERS. (1987). *Steel grades for the manufacture of welded pipes, resistant to cracking under the influence of hydrogen sulphide*. Document of the working group on the metallurgical industry of Franco-Soviet cooperation (p. 51).

Yamada, K. et al. (1983). Influence of metallurgical factors on HIC of high strength ERW line pipe for sour gas service. In: *International Conference on Technology and Applications of HSLA Steels, Philadelphia, Pensylvania* (pp. 835–842). October 3–6, 1983..

Chapter 3
Central Chemical and Structural Segregation Heterogeneity in Continuously Cast Slabs

Continuously cast slab serves as a billet for the production of high-quality plate and coiled stock for pipes. The initial quality of slab largely determines the ability to meet the requirements on surface quality, contamination with non-metallic inclusions, impact toughness, crack resistance in H_2S-containing media, continuity, etc. The quality of continuously cast billet is formed at the stages of liquid steel treatment in the ladle, in the vacuum unit, in the ladle-furnace unit, in the intermediate ladle, and upon continuous casting. Thereby, the problems of ensuring the chemical composition within specified limits, the low content of harmful impurities (sulfur, phosphorus) and gases (nitrogen, hydrogen, oxygen), cleanliness from non-metallic inclusions, minimization of surface and internal defects of the slab, etc., are solved.

The resistance of steel to fracture by the mechanisms of hydrogen-induced cracking and sulfide stress cracking in hydrogen sulfide-containing media to a large extent depends on the distribution and morphology of non-metallic inclusions and on the degree of the central segregation heterogeneity formed in slab upon solidification.

Chemical segregation is an essential part of the melt solidification process upon continuous casting of steel. The maximum segregation heterogeneity is formed in the axial region of the cast billet and is retained upon the subsequent rolling. In most cases, a certain level of segregation heterogeneity is permitted. However, for hydrogen sulfide-resistant steels, this level should be substantially limited and carefully controlled. Below, we consider the factors affecting the quality of a continuously cast billet and some methods for reducing the degree of the central segregation heterogeneity of slabs.

© Springer Nature Switzerland AG 2019
I. Shabalov et al., *Pipeline Steels for Sour Service*, Topics in Mining, Metallurgy and Materials Engineering, https://doi.org/10.1007/978-3-030-00647-1_3

3.1 Macrostructure of Continuously Cast Slab

Transition of melt from liquid to solid state is accompanied by crystallization of steel and solidification of the ingot. The precipitation of individual crystals from the melt and the formation of structure zones of the ingot are often called crystallization. Solidification means the process of increasing the quantity of solid phase and, correspondingly, decreasing the volume of liquid phase in the continuously cast billet.

The crystallization process plays an important role in the formation of the structure and properties of steel and is one of the leading factors for obtaining the necessary properties of metal products. The following stages can be distinguished in it:

- nucleation of crystallization centers;
- formation of crystals in the ingot;
- segregation of impurities and development of dendritic heterogeneity;
- convective motion in the volume of the solidifying melt;
- occurrence of zonal chemical and structural heterogeneity in the solidifying melt.

The crystallization character largely determines the shape, orientation, and structure of the crystals and the motion of the liquid phase through the network of growing crystals. Under normal conditions, when the continuously cast metal is solidified, crystals of a branched, dendritic shape are formed (Fig. 3.1).

The direction of the major dendrite axis coincides with certain crystallographic orientations of the lattice. The dendritic structure affects the development of segregation processes in the solidifying ingot, grain size, and the mechanical properties both in the cast and deformed states.

The formation of various structure zones at the stage of solidification is due to the changes in the physical and physicochemical properties (chemical composition of the metal, temperature, etc.) of the ingot upon crystallization. The formation of structure zones and the development of segregation processes in continuously cast ingots are determined by the following factors:

- cooling rate of liquid steel from the periphery to the axis of the ingot and the temperature gradient in the melt at the solidification front;
- development of a wide segregation layer with a high concentration of impurities and a low solidification temperature promoting the nucleation of endogene crystallization centers ahead the crystallization front;
- convective (upward and downward) and gravitational (downward) flows in liquid steel and the non-uniform crystallization caused by such flows.

In steel ingots, three main zones can be distinguished:

- outer crust zone of fine misoriented "frozen" crystals;
- zone of columnar oriented dendritic crystals;
- zone of misoriented equiaxed crystals.

An example of the macrostructure of continuously cast ingot from low-carbon low-alloy microalloyed pipe steel is shown in Fig. 3.2.

Fig. 3.1 Scheme of the dendritic crystallization of metal: 1—seed; 2—needle crystal; 3—vertex shape, which does not change in time; 4—dendrite with lateral branches

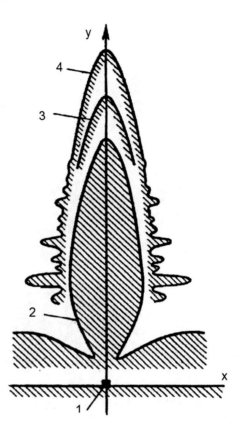

When liquid steel comes into contact with the water-cooled wall of the CCM mold, many crystallization centers arise in the contacting layer due to a large temperature gradient. As a result, a zone of fine misoriented crystals is formed on the surface of the cast billet. The presence of a great number of nuclei substantially suppresses the crystal growth. Therefore, the zone of crust ("frozen") crystals consists of a large number of fine misoriented crystals. The length of this zone in cast slabs is 2–4 mm. An example of a typical zone of crust crystals in continuously cast slabs is shown in Fig. 3.3.

The resulting outer shell of crust crystals increases the resistance to heat transfer from the liquid phase to the cooled wall of the mold. This leads to a decrease in the temperature gradient in the solidifying ingot. A decrease in the intensity of heat removal changes the character of the solidification process. The formation of the second structure zone begins in the slab. Crystals grow, penetrating deep into the liquid metal. They are elongated in shape. Usually, they are called columnar crystals, and the zone occupied by such crystals is a zone of directional crystallization (transcrystallite zone). The columnar crystals can penetrate the entire thickness of the metal to the thermal axis (Fig. 3.4).

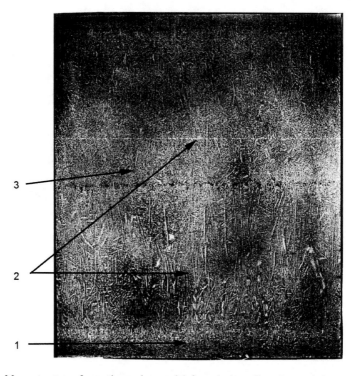

Fig. 3.2 Macrostructure of a continuously cast slab from the low-alloy pipe steel: 1—zone of crust "frozen" crystals; 2—zone of columnar dendrites; 3—zone of misoriented dendrites (Belyj et al. 2005)

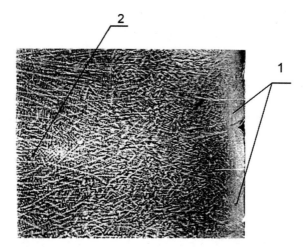

Fig. 3.3 Zone of crust crystals (1) and columnar dendrites (2) (Belyj et al. 2005)

Fig. 3.4 Zone of directional columnar crystallization (Belyj et al. 2005)

With time, the temperature of the liquid phase retained in the ingot decreases and reaches the solidification temperature. The attainment of the liquidus temperature in the central region of the ingot and the presence of the crystallization nuclei such as solid non-metallic inclusions in the liquid phase lead to the formation of a zone of misoriented crystals whose axes are oriented randomly (Fig. 3.5a). This is the third zone of the solidifying ingot. The equiaxed crystals begin to nucleate in the melt at the interface between the liquid and solid phases. This is explained by the fact that the central liquid region is supercooled due to the solute segregation at the growing interphase interface is such a way that the particles causing the formation of equiaxed crystals can be nucleated in the supercooled zone. In addition, the equiaxed crystals can be formed from partially melted dendrites as the melt temperature fluctuates upon the growth of the columnar zone. A sufficient overheating provides favorable conditions in a liquid well for the formation of isolated crystals due to submelting and breaking off the dendrite branches.

In some cases, a zone of fine globular crystals without any clear dendritic structure is formed along with the misoriented crystals in the axial zone of the ingot (Fig. 3.5b). Such crystals can nucleate on the mold wall or on the cooled surface of the melt and grow with the formation of necks. Then such neck-like crystals are separated from the nucleation site before the solid shell is formed, finally precipitated, and accumulated to form the zone of equiaxed structure.

The macrostructure of the ingots cast in continuous casting machines is characterized by a number of specific features caused by the unit design. The main

(a)   **(b)**

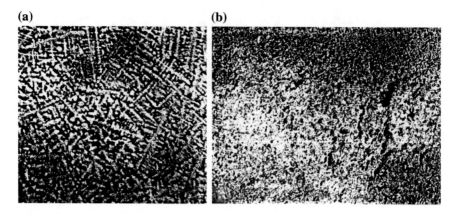

Fig. 3.5 Regions of misoriented equiaxed crystals (**a**) and globular crystals (**b**) in the axial zone of the ingot (Belyj et al. 2005)

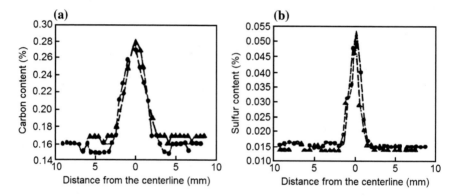

Fig. 3.6 Contents of carbon (**a**) and sulfur (**b**) in the axial zone of continuously cast slab

macrostructure characteristic of the slab cast by curved CCM is its asymmetry. The different lengths of the zones of columnar crystals on opposite sides of the slab are the result of the action of physicochemical regularities (the effect of gravitation flows of liquid steel, sedimentation of broken dendrite branches), and it is impossible to completely eliminate this difference. To an extent, it can be reduced by decreasing temperature of the cast metal and by decreasing casting speed.

The central segregation is most pronounced in the continuously cast slabs. The axial segregation is present, as a rule, in the thermal center of the slab at a distance of 0.5–3.0 mm from its geometric axis. The content of segregating elements in the segregation zone can exceed that in the initial metal by a factor of more than 3–5 (Fig. 3.6).

Substantial fluctuations of the degree of segregation in slabs of the same thickness show that the character of crystallization cannot be considered as the only cause of the chemical heterogeneity of the ingot. The factors affecting the development of zonal heterogeneity include the redistribution of segregates and their concentration in local zones due to mass transfer processes such as forced and natural metal motion caused by hydrodynamic and convective flows, the forces of capillary mass transfer, and the bulging phenomenon.

The main factors determining the solidification processes and heterogeneity of the ingot are intense jet-circulation flows caused in the mold by the hydrodynamic action of the metal jet from the intermediate ladle. Upon subsequent movement, the ingot falls into the zone of damping the hydrodynamic flows and the formation of still weak convective flows under conditions of an increased concentration of impurities in the liquid melt ahead the solidification front. Judging by the nature of impurity distribution, the natural flows begin to form at the end of the zone of positive segregation, when the crust is 35–40 mm thick. As soon as the crystals clamp together and form a motionless "skeleton," a network of channels, in which the liquid metal is enriched with segregating impurities, appears at the solidification front.

The natural convective flows are less developed and rapid at a wide zone of misoriented dendrites and at limited extension of the zone of transcrystals. The efficiency of the enrichment of residual melt with impurities is reduced, the segregates are distributed over a larger volume, and the zonal heterogeneity is less pronounced. As the zone of columnar dendrites becomes smaller by a factor of 2–3, the axial segregation decreases by a factor of 1.8–2.0.

The main conclusion about the mechanism of chemical heterogeneity realization in a continuously cast slab is as follows. The segregation phenomena are caused by the development of forced and natural convective flows enriching the residual melt with impurities, by the morphology of the solidification front, and by the filtration processes, which are enhanced by mechanical action on the billet shell and promote the impurity concentration in the axial zone.

The maximum number of non-metallic inclusions, mainly sulfides concentrated along the grain boundaries, is observed in the axial zone of the ingot. In some cases, a large amount of silicates and aluminates is observed in the axial zone (Fig. 3.7).

3.2 Methods of the Study of Central Segregation Heterogeneity

With advancing technology of continuous casting of steel, the methods were developed for identification and estimation of the internal structure defects and segregation heterogeneity of continuously cast billets in industrial conditions. Such methods include the method of Baumann sulfur prints and etching of macrotemplates in solutions of various acids. Macrotemplates taken from slabs are used for the examination. The macrotemplates are cut in such a way that the section to be examined is separated from the cutting site by a distance excluding the effect of cutting conditions (heating by autogenous cutting, crushing by the press, saw, etc.). The macrotemplate section is prepared by cold machining (gouging, grinding, etc.). Macrotemplates are etched

(a)   **(b)**

(c)

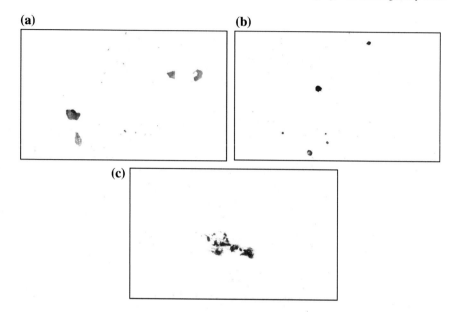

Fig. 3.7 Non-metallic inclusions in the axial zone of a continuously cast slab: a—sulfides; b—silicates; c—aluminates (Belyj et al. 2005)

up to the identification of macrostructure defects and then scored by comparing the natural view of templates or prints with schematic standards.

The method for taking and estimation of sulfur prints is sensitive to sulfur content and can be used to evaluate the internal structure of slabs with a sulfur content of more than 0.005%, although serious malfunctions in the CCM operation can be determined from prints at sulfur content in the steel of $\geq 0.002\%$.

To determine the macrostructure and segregation in the steels containing $\leq 0.005\%$ sulfur, etching in the following reagents is used: 10% aqueous solution of ammonium persulfate $(NH_4)_2S_2O_8$; electrolyte; 50% aqueous HCl solution at a temperature of 60–80 °C; aqueous solution (10% HCl and 1–1.5% H_2O_2); aqueous solution of 1.5% picric acid and 0.5% $CuCl_2$.

Each of the above reagents is not universal and has individual etching characteristics depending on the carbon and phosphorus contents of the steel. Etching with picric acid requires very careful preparation of the surface. Hydrochloric acid and mixed chloride etching substances are well suited for detecting dendritic structure, but not so useful for obtaining clear prints of the axial zone. At the same time, etching with hydrochloric acid and picric acid requires a vent hood because of the possibility of harmful vapor release. Ammonium persulfate gives the results of a medium level for etching specimens containing more than 0.04–0.05% carbon, in addition, it is a strong oxidizer. Etching with picric acid gives good results for the steels with higher phosphorus concentrations and requires careful surface treatment at low phosphorus levels. The most widely used method is warm etching of macro templates in hydrochloric acid, 50% aqueous solution.

The specific choice of the etching method is determined by a number of criteria such as:

- obtaining prints with a pronounced segregation area;
- etching duration;
- usability;
- availability of reagents and related equipment;
- required quality of surface preparation;
- suitability for the study of steels of different chemical compositions;
- compliance with safety standards.

In the production practice, the Russian industry standard OST 14-4-73 and the document "Mannesmann rating system for internal defects in continuously cast slabs" are used to evaluate and control the macrostructure of continuously cast billets.

According to OST 14-4-73, the following types of defects are distinguished:

- axial looseness;
- axial chemical heterogeneity;
- axial cracks;
- cracks and segregation bands perpendicular to the narrow and wide faces or angular;
- nest cracks;
- point heterogeneity;
- surface carburization.

The reference scales include four numbers (from 1 to 4), and the macrostructure free of defects is estimated by a number of 0. The estimation by fractional numbers (0.5, 1.5, etc.) is permitted. Figure 3.8 shows the scale for the estimation of chemical heterogeneity characterized by the standard as enrichment or depletion of the axial zone by impurities. Such effects are revealed by the change in the color of the sulfur prints or by the zone of stronger or weaker etchability than that of the base metal. The number of chemical heterogeneity increases with increasing brightness or darkness of the points and their sizes.

According to the Mannesmann system of five classes, the internal defects of continuously cast slabs are divided into the following types:

- transverse internal cracks;
- corner cracks;
- center segregation and core unsoundness;
- spot-shaped inclusions;
- cloud-shaped inclusions;
- narrow side cracks;
- longitudinal internal cracks.

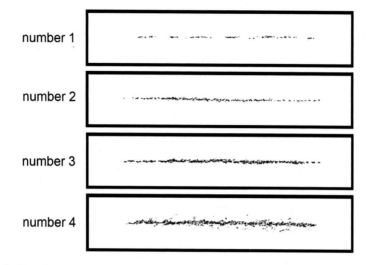

Fig. 3.8 Classification of the axial chemical heterogeneity according to the OST 14-4-73

In slabs of low-carbon low-alloy pipe steels, depending on the content of strongly segregating elements such as carbon, manganese, sulfur, and phosphorus, the central segregation heterogeneity predominantly ranges between numbers of 0 and 3 according to OST 14-4-73 and from class 1 to class 3 according to the Mannesmann classification system (Fig. 3.9).

The central segregation chemical and structural heterogeneity over thickness of slab or rolled product in laboratory conditions is evaluated by the following methods:

- determination of the content of chemical elements over thickness of slabs and plates;
- metallographic examination of macro- and microstructure;
- microhardness measurement;
- by determining the difference between the contents of chemical elements or hardness in the axial zone and the base metal;
- by using the segregation chemical and structural heterogeneity coefficients determined as the ratios between the element contents or hardnesses in the axial zone and in the base metal.

3.3 Factors Affecting the Centerline Segregation

Upon continuous casting, the macrostructure formation and, as a consequence, the degree of the central segregation chemical and structural heterogeneity of slabs are affected by a number of factors, the following of which are dominant (Belyj et al. 2005):

(a)   **(b)**

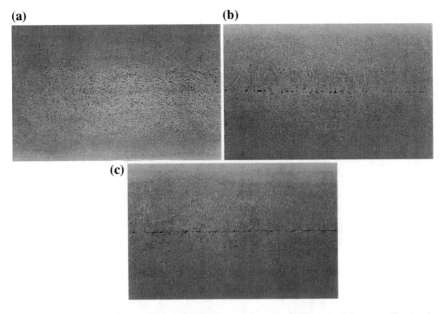

(c)

Fig. 3.9 Center segregation and core unsoundness of continuously cast slabs according to the Mannesmann classification of internal defects: **a** class 1, **b** class 2, **c** class 3

- superheating of liquid steel over the liquidus temperature;
- steel casting speed;
- adjustment of the roller guide of CCM;
- size of the mold cross section;
- chemical composition of the metal;
- secondary cooling regime;
- CCM design features.

Overheating of the liquid metal relative to the liquidus temperature in the intermediate ladle upon continuous casting is one of the most significant factors determining the character of the macrostructure formed in the slab. For iron–carbon alloys, there is a linear relationship between the length of columnar crystals in the cross section of the ingot and overheating relative to the liquidus temperature. A decrease in overheating reduces the zone of columnar crystals and the degree of axial segregation of chemical elements. Small overheating promotes the expansion of the zone of misoriented crystals and dispersion of shrinkage and segregation phenomena. For example, a decrease in the casting temperature from 1545 °C to 1515–1520 °C allows one to reduce the number of templates with axial segregation above a number of 2 by a factor of more than two. With increasing temperature of the cast metal, an increase in the axial segregation of sulfur and phosphorus is primarily observed. It is desirable that the overheating upon continuous casting of steel does not exceed 15 °C. In this case, there are no internal defects such as axial, perpendicular, and longitudinal cracks, a fine-grained equiaxed structure is formed, the distribution of non-metallic

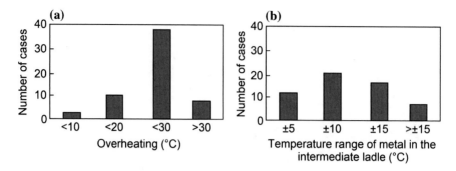

Fig. 3.10 Distribution of the overheating (**a**) and the temperature range of steel in the intermediate ladle (**b**) provided by casting technology for a number of manufacturers of continuously cast slabs (total 55 cases) (Belyj et al. 2005)

inclusions is uniform, and the central porosity is small. With increasing overheating, the number of surface defects and internal cracks increases. The number of internal slab defects increases by a factor of 1.5–2 with increasing overheating from 0–5 to 20–25 °C.

The presence of a developed zone of equiaxed crystals substantially reduces the degree of axial segregation of elements. A proportional relationship is observed between the increase in superheating of steel and the narrowing of the equiaxial zone. As the overheating increases from 2 to 20 °C, the extension of the zone of equiaxed crystals is reduced from 40 to 15%.

The effect of overheating on the ingot macrostructure is particularly pronounced as it is below 15 °C. At a slight overheating (up to 5 °C), a non-oriented structure is formed almost in the entire volume of the ingot. When the steel is overheated by more than 30 °C above the liquidus temperature, the macrostructure changes virtually end, and the relationships between the structure zones become stable.

At the same time, at small degrees of steel overheating, troubles arise in the casting process, which lead to the stricture of the channels of the immersed casting nozzles, metering nozzles of the intermediate ladle, and ladle sliding gates, to the formation of a "crust" on the surface of the liquid metal in the mold, etc.

At present, steel is cast mainly at the lower temperature limit, at which the casting technology is not violated. For example, the optimum metal temperature in the intermediate ladle for casting of low-alloy steel slabs of 250 × 1550 … 1850 and 300 × 1850 mm in cross section is 1534–1545 °C. Most steel producers perform continuous casting at overheating of less than 30 °C relative to the liquidus temperature and carefully adjust the metal temperature in the intermediate ladle (Fig. 3.10).

The axial segregation has the greatest effect on the consumer properties of critical duty microalloyed peritectic manganese steels. Steel casting at the lowest possible temperature with the elimination of the drawbacks inherent in casting at small overheating is one of the principal ways to improve the quality of cast metal.

The effect of a number of technological factors on the development of the chemical and structural heterogeneity in continuously cast slabs in double-strand curved-mold

Table 3.1 Chemical composition of the steels (Belyj et al. 2005)

Steel	Elements (wt%), within or maximum							
	C	Mn	Si	S	P	Al	V	Nb
S355J2	0.18–0.21	1.40–1.60	0.15–0.40	0.006	0.020	0.020–0.050	–	0.020–0.040
09G2FB	0.08–0.11	1.50–1.70					0.050–0.070	

Table 3.2 Effect of the overheating relative to the liquidus temperature upon continuous casting of the 09G2FB steel on the structure zone extensions of the slabs of 250 × 1850 mm in cross section (Belyj et al. 2005)

Average overheating relative to the liquidus temperature (°C)	Extension of structure zones (%)			Number of experimental points
	Crust	Columnar	Equiaxed	
10	7.1	45.6	47.3	5
15	6.4	54.3	39.3	9
20	6.0	67.0	27.0	12
25	5.7	68.6	25.7	44
30	3.2	72.3	24.5	48
36	2.4	73.8	23.8	17

continuous casting machines was studied in (Belyj et al. 2005). The S355J2 structural steel and the 09G2FB pipe steel (Table 3.1) were cast into the mold of 250 × 1850 mm in cross section. It is noted that the development of axial segregation of chemical elements is directly related to the crystal structure of the continuously cast billet, i.e., to the relationship between the extensions of the structure zones of the ingot.

The studies were performed on the templates taken from the third or fourth slabs from one or both strands. The sizes of the zones of crust "frozen" crystals, columnar dendrites, and misoriented dendrites were determined over the template area. Upon the slab macrostructure examination, the overheating of the metal relative to the liquidus temperature, the linear rate of steel casting, and the slab thickness were taken into account. The effect of steel overheating relative to the liquidus temperature was estimated on heats from the 09G2FB steel cast into a mold of 250 × 1850 mm in cross section at a rate of 0.7–0.8 m/min. The data on the effects of the metal temperature in the intermediate ladle upon continuous casting and the overheating on the dimensions of the structure zones of the slab are given in Table 3.2 and shown in Fig. 3.11.

As follows from the data given in Table 3.2 and Fig. 3.11, an increase in the metal temperature in the intermediate ladle and, correspondingly, an increase in the overheating of the steel upon continuous casting promote an increase in the extension of the zone of directional crystallization (zones of columnar crystals) with simultaneous decrease in the sizes of the zones of crust crystals and misoriented dendrites. The structure is most affected by the smallest overheating (up to 20 °C), while a further increase in overheating weakens its effect on the structure, and the change in the relationship between of structure zone extensions is insignificant. In this case, the relationship between the zones becomes stable: The columnar crystal zone

Fig. 3.11 Extension of
structure zones in the
09G2FB steel continuously
cast slabs of 250 × 1850 mm
in cross-sectional size as a
function of the metal
temperature in the
intermediate ladle: I—zone
of equiaxed crystals;
II—zone of columnar
dendrites; III—zone of crust
crystals (Belyj et al. 2005)

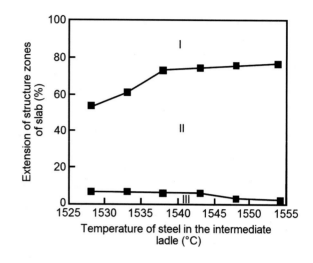

Table 3.3 Dependence of the
axial chemical heterogeneity
on steel overheating relative
to the liquidus temperature
(Belyj et al. 2005)

Overheating (°C)	Average number of the axial chemical heterogeneity according to the OST 14-4-73	
	09G2FB	S355J2
15 (12–17)	1.20	1.45
20 (18–22)	1.25	1.50
25 (23–27)	1.40	1.55
30 (28–33)	1.55	2.00
36 (34–40)	1.70	2.10

occupies 70–75%, and the sum of the equiaxed and crust zones is about 25–30%. At an overheating of 10–15 °C, the zone of columnar crystals is substantially reduced (to 45.6%) with simultaneous growth of the crust zone and zone of misoriented dendrites. The zone of globular crystals appears as the steel is overheated by <10 °C.

The effect of overheating on the macrostructure imperfection, in particular, on the development of axial segregation of chemical elements is more complicated. The data given in Table 3.3 demonstrate the effect of overheating on the average number of the axial chemical heterogeneity of the 09G2FB and S355J2 steel slabs.

It is seen that, as an average overheating temperature increases from 15 to 36 °C, the average number of the axial segregation of elements increases from 1.2 to 1.7 for the 09G2FB steel and from 1.45 to 2.1 for the S355J2 steel. The axial chemical heterogeneity in the S355J2 steel slabs is higher than that is in the 09G2FB steel slabs. This can be explained by different carbon contents: The axial segregation at a carbon content of 0.18-0.21% is higher by a factor of 1.2–1.3 than that is at a carbon content of 0.08–0.11%.

The formation of structure zones and the development of internal defects are substantially affected by the steel casting speed. The sudden change in the casting speed observed at so-called transitional regimes detrimentally affects the quality of the cast metal. Transitional regimes are characterized by simultaneous deviation of temperature and casting speed from optimal magnitudes. In this case, the following parameters change:

- the conditions of heat removal from the solidifying ingot to the water-cooled wall of the mold;
- hydrodynamic pattern of the distribution of submerged jets in a liquid pool and the conditions for the floating-up of non-metallic inclusions into the covering slag;
- viscosity parameters of the slag-forming mixture and the conditions of its operation in the gap between the mold wall and the ingot crust;
- conditions for evolution of gases, primarily hydrogen, from the solidifying ingot into the gas gap;
- thermo-stressed state of the ingot, including that in the straightening zone of CCM.

With increasing casting speed, the number and extension of internal cracks increase. In particular, an increase in the casting speed from 0.7 to 0.9 m/min leads to an increase in the number of the parameter of internal cracks and their length, on average, by a factor of 2.5. An increase in casting speed also affects the axial segregation of chemical elements. With increasing casting speed from 0.7 to 0.9 m/min, the axial porosity of the continuously cast billet increases to a number of 1. Therefore, most manufacturers of critical duty steels, for example, for the production of HIC-resistant gas pipelines use a casting speed of less than 0.9 m/min. A short-term reduction in casting speed leads to the formation of internal cracks (Fig. 3.12).

An improvement in the macrostructure upon a decrease in the casting speed is associated with increasing joint angle of the crystallization fronts, decreasing ratio between the liquid pool depth and the slab thickness, increasing fraction of the two-phase region, averaging temperature over volume, and decreasing temperature gradient. All this leads to an abrupt decrease in the quantity of longitudinal cracks and the degree of axial chemical heterogeneity in the slabs of 240 × 1850 mm, 240 × 1550 mm, and 240 × 1710 mm in cross sections with decreasing casting speed from 0.4 to 0.1 m/min.

The development of axial segregation of elements substantially depends on the adjustment of the roller guide of CCM. It is necessary to avoid displacement of the rollers in the secondary cooling section since this can lead to local enrichment of the residual liquid phase with segregating elements. Bulging of the slab in the sites where contact with the roller is lost as a result of its beating can lead to a periodically high HIC level. The deflection of the rollers is also a serious factor affecting the development of the axial segregation of chemical elements and the generation of axial cracks. At a substantial deviation in the setting of the roller guide from the optimal one determined by the design of the continuous casting machine, the factor of the deviation of the rollers from the initial positions prevails upon the formation of axial defects of the "delamination" type. The bulging of the solidifying slab casing between the rollers of the guide is caused by the ferrostatic pressure of the column of liquid metal. Bulging of the slab in the inter-roller space alternates with the reduction of the billet by rollers, and this leads to the formation of periodically compressive

Fig. 3.12 Defects resulting
from a short-term abrupt
decrease in the casting
speed: 1—nest crack;
2—axial crack

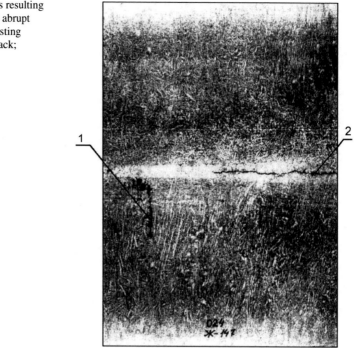

and tensile stresses and deformations. The cyclic change in "compressive-tensile" stresses leads to the formation of the flows of metal enriched with segregates along the solidification front. The presence of the zone of directional crystallization promotes the development of the enrichment with segregates in the axial zone of the solidifying slab. Reducing bulging of the crust of the ingot substantially reduces the concentration of segregates in the axial zone of the slab.

A decrease in the diameter of rollers allows one to shorten the distance between them, correspondingly, to reduce the slab bulging, and, thus, to avoid the formation of defects in the axial zone of the ingot. Reducing the bending of the rollers positively affects the density of the axial zone.

The dimensions of a continuously cast slab, primarily its thickness, affect the macrostructure and the development of axial segregation. Such effect is dual, and it has no general clear pattern. This is explained by the fact that, on the one hand, the degree of segregation in the axial zone decreases with increasing cross-sectional area of the billet due to the conventional reduction in the casting speed of steel with increasing cross-sectional area of the billet. On the other hand, a decrease in the cross-sectional area of the ingot leads to the macrostructure improvement, which is expressed by a reduction in the zone of directional crystallization. With decreasing slab thickness, the length of the liquid pool and the degree of bulging of the cast billet are reduced. This reduces the number of defects in the axial zone of the ingot. The reduction in the thickness of the 3sp, 17G1S-U, and 09G2S steel billets allows one to

Fig. 3.13 Width of the
columnar crystal zone as a
function of the 09G2FB and
10G2FB steel slab thickness
at different metal
temperatures in the
intermediate ladle:
1–250 mm; 2–300 mm
(Belyj et al. 2005)

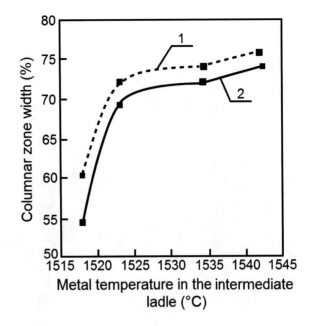

reduce the number of billets with axial cracks by 39% and to reduce the coefficient
of carbon segregation in the axial zone from 1.65 to 1.30.

The effect of the thickness of a continuously cast billet on the columnar crystal
zone width in the 09G2FB and 10G2FB steel plates was studied in (Belyj et al. 2005)
(Fig. 3.13).

The width of the columnar crystal zone in a slab 250 mm thick is, on average,
higher by 3% than that is in a slab 300 mm thick, irrespective of the overheating
temperature of the steel. Correspondingly, the degree of the development of axial
segregation in a slab 250 mm thick is also more pronounced (Fig. 3.14). In going
from a slab 250 mm thick to a slab 300 mm thick, it is possible to reduce the axial
segregation and the number of defects such as internal cracks and axial looseness with
simultaneous expansion of the zone of equiaxed crystals. In general, for each specific
case of the CCM roller guide design and the adopted continuous casting technology,
the question of the relationship between the quality of the internal structure and the
cross-sectional size of the ingot is a subject of a special study.

3.4 Methods of Decreasing the Degree of Central Segregation

In the continuously cast slabs and rolled products from low-alloy steels, various
types of the segregation of chemical elements are observed. The most negative effect
is caused by the segregation in the zone adjacent to the centerline of the slab, i.e.,

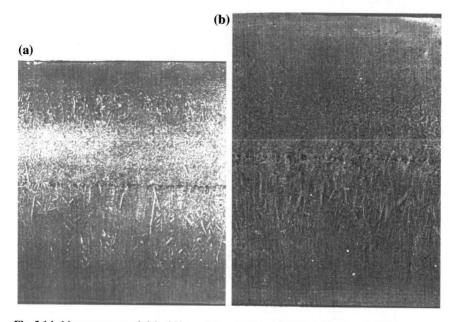

Fig. 3.14 Macrostructure of slabs 250 mm (**a**) and 300 mm (**b**) thick (Belyj et al. 2005)

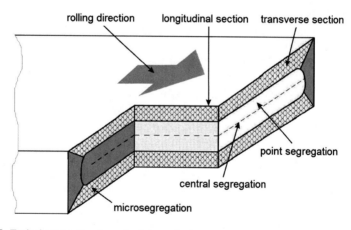

Fig. 3.15 Typical segregation types in plate products

central segregation (Fig. 3.15). A highly developed central segregation zone can cause undesirable phenomena such as discontinuity in the finished rolled steel, a decrease in the resistance to cracking in H_2S- and CO_2-containing media, deterioration of weldability.

Taking into account the great effect of segregation heterogeneity on the properties of plate metal, a number of technological measures aimed at reducing it in continuously cast slabs have been developed:

- electromagnetic stirring of the melt in the secondary cooling zone;
- vibrational and impulsive action on the crystallizing metal;
- "soft" reduction;
- introduction of a solid phase into the melt in the mold of the continuous casting machine.

Such methods to some extent affect the heat and mass transfer, reduce the chemical heterogeneity of the metal, and refine the grains of the cast structure. An effective way to reduce the intensity of the development of central segregation heterogeneity and its consequences is the optimization of the chemical composition of the cast steel.

3.4.1 Technological Methods for Improving the Quality of the Slab Macrostructure

Electromagnetic Stirring EMS allows one to affect the macrostructure and quality of a continuously cast ingot. Electromagnetic stirring units are installed in two positions along the technological line of slab CCM: on the mold or directly below it and in the secondary cooling zone (Fig. 3.16).

When using this method, the positive effect of forced stirring of the liquid metal on the formation of the structure of a continuously cast ingot is noted. In this case, heat transfer from the liquid core of the ingot to the crystallization front is enhanced, and non-metallic inclusions and dissolved gases float up. As a result, the macrostructure quality is improved, and, to some extent, the development of axial segregation of elements is suppressed.

In the process of the industrial development of EMS, along with its advantages, certain drawbacks of the method were also found, namely it was not always possible to eliminate axial segregation since the use of several inductors increases the cost of the process.

To improve the macrostructure quality, it is proposed to place the EMS unit as close as possible to the mold. At the same time, as a result of stirring, the dendrite branches break off and melt, the heat removal to the solid crust of the ingot increases, and, correspondingly, the overheating of steel relative to the liquidus temperature decreases.

One of the most important advantages of EMS is that this method promotes the formation of equiaxed crystals upon solidification of continuously cast ingot and leads to a more uniform distribution of non-metallic inclusions.

All EMS methods to some extent reduce axial segregation due to the creation of a large zone of equiaxed crystals. The method of electromagnetic stirring is further developed, but it has a number of disadvantages, including high cost of devices, significant energy consumption, and insufficient efficiency of the action on slabs of large cross sections.

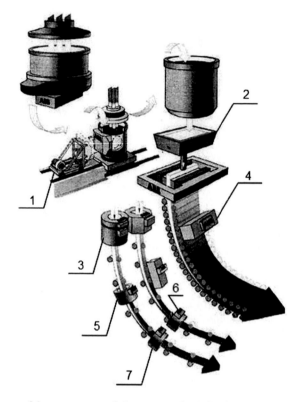

Fig. 3.16 Scheme of the arrangement of electromagnetic stirring (EMS) systems and their design on the mold of slab CCM: 1—inductor on the steel ladle; 2—electromagnetic brake device; 3—EMS under the mold; 4—slab EMS; 5—EMS; 6—electromagnetic inductor; 7—final EMS (Belyj et al. 2005)

The applied methods of vibrational processing, ultrasonic treatment, and hydroimpulse action are based on the excitation of oscillations in metal at the crystallization stage. The application of elastic oscillations periodically increases or sharply reduces pressure on the solidification boundary, and thus affects the segregation and thermal processes of ingot formation. There are two methods of imposing elastic vibrations such as vibration of metal melts at a frequency of ≤ 25–50 Hz (low-frequency vibration) and ultrasonic vibration at a frequency of up to 20,000 Hz.

The Method of Vibroprocessing of Liquid Steel is realized with the help of a vibrator creating directional vibrations in the vertical plane or imposing low-frequency vibration perpendicular to the wide faces of the billet. Due to vibration, the quality of the metal is improved, in particular, the degree of central segregation and the development of axial porosity are reduced, and the shell of the continuously cast ingot is formed more uniformly. The vibroprocessing can reduce the size of the columnar zone and substantially increase the globular zone.

For Ultrasonic Treatment of a Liquid Metal Jet, a unit is used, which is a heat exchanger between the ladle and the CCM mold. Upon such treatment, the crystallization rate increases by 35–50%, the liquid pool length is reduced by 25%, the width of the columnar dendrite zone decreases, and the equiaxed crystal zone is simultaneously expanded from 32 to 70% of the ingot area.

The Method of Gas-Pulse Treatment considered in allows one to provide a spatial character of solidification and to obtain fine grains in the ingot. The method of pulsating mixing of the melt is based on the processing of liquid bath by a pulsating gas jet, which is fed from a submerged pipe and promotes the formation of directed circulation flows and zones of increased turbulence. The pulsating regions of the liquid flows provide a high intensity of volume stirring and create favorable conditions for the development of cavitation flows, which contribute to the formation of additional crystallization centers.

The disadvantage of the above methods is the high energy consumption upon their application. In addition, it is quite difficult to avoid the impact of vibration on equipment.

The Method of Electrohydropulse Treatment employs the generation of elastic vibrations caused by an electric discharge in liquid metal. Since the required specific power consumption is not too high, the method is suitable for processing large metal volumes and can be used for the continuous casting of steel. More complex devices are required for its use. The powerful pulses generated by an electric discharge are introduced into liquid center of the forming ingot and destroy the crystallization front. The melt is effectively mixed, and the conditions for the formation of additional crystallization nuclei result from the development of cavitation phenomena in the liquid center.

The principle of electrohydropulse impact on metal upon continuous casting is as follows. The electric energy accumulated in a capacitor bank is released in the interelectrode gap of the electric-discharge generator of elastic oscillations in the form of a plasma channel of high temperature and pressure. Upon high-speed expansion of the plasma channel in water, shock waves are generated in the electric-discharge generator of elastic oscillations and through the membrane transmit elastic vibrations to the forming ingot.

The use of electrohydropulse treatment substantially improves the quality of continuously cast metal. For example, in the experimental metal, the central segregation decreased by a number of 1–1.5, the axial looseness decreased by a number of 0.5–1, the fineness of the dendritic structure increased by 25–30%, the number of primary dendrites per 1 cm^2 of the template area increased from 18 to 36, the number of globular crystals increased from 24 to 42, the segregation coefficient of the elements decreased, and the quality indicators of rolled steel improved.

However, all these methods to date have not found wide practical application and are at the stage of pilot-industrial development.

In contrast to this, *the system of "soft" reduction* has been widely used since the 1990s, after the introduction of new designs of CCM roller guide. The system provides the reduction of the continuously cast billet in the strand upon its production

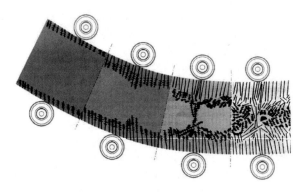

by compressive deformation of a partially solidified slab in order to remove the segregation liquid (Fig. 3.17).

When using the "soft" reduction method, it is necessary to precisely determine the position of the reduction force application with allowance for the relationship between the amounts of liquid and solid phases at the time of reduction. The maximum effect of the suppression of axial segregation is achieved at minimum divergence of technological parameters of casting, which is possible at careful control of the whole technological cycle of steel production, complex automation, and computerization of processes.

In modern units, an automatic adjustment is made upon continuous casting of steel at a rapid change in the slab thickness. The reduction of the slab occurs in a horizontal section of the roller guide before the final solidification in the presence of a two-phase zone in the axial part of the slab. The control system automatically tracks the metal temperature in the intermediate ladle and the casting speed and on the basis of these data calculates the area, at which the crystallization of the billet is completed. Based on the calculations of the fraction of solid phase in the axial region of the slab along the length of the roller guide, the positions of start and end of the zone of "soft" reduction are determined. The degree of reduction at each enterprise is determined by the results of industrial experiments, and usually it is from 2 to 6 mm over thickness or up to 1.5 mm per 1 m of length. The effect of a "soft" reduction on the degree of carbon segregation in a continuously cast slab is shown in Fig. 3.18.

Despite the widespread use of the "soft" reduction method, its efficiency in some cases was shown to be insufficiently high even at optimal process conditions at an exact calculation of the position, size, and character of the reduction.

One of the ways of improving the slab macrostructure and decreasing the axial segregation is the introduction of a solid phase such as microcoolers (fine inoculators) or consumable macrocoolers into the liquid metal upon continuous casting.

The introduction of inoculators allows one to decrease the overheating of liquid steel and the temperature gradient over the cross section of the slab and to increase the number of heterogeneous crystallization centers. All this beneficially affects the structure of the metal, reducing the length of the zone of directional crystallization as well as the chemical and structural heterogeneity. The use of microcoolers decreases

Fig. 3.18 Effect of a "soft" reduction on the coefficient of carbon segregation (Belyj et al. 2005)

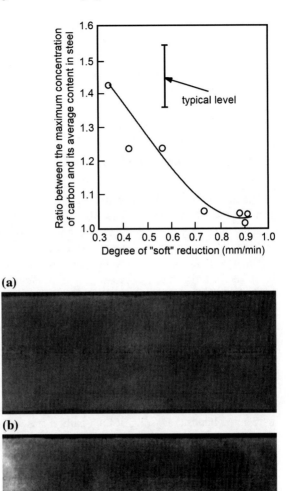

Fig. 3.19 Macrostructure of slabs continuously cast by the standard technology (**a**) and with the introduction of strip (**b**)

(a)

(b)

the central segregation number from 2.5–3 to 1.5–2 and a quantity of transverse and longitudinal cracks. However, this method has a number of substantial drawbacks such as an increase in the content of non-metallic inclusions in the metal, the lack of reliable devices for the introduction of microcoolers, a non-uniform distribution of microcoolers over the volume of the melt.

Use of consumable macrocoolers, which include various types of strips, plates, and wires, is performed by their introduction into the liquid pool of the CCM mold. As a result, the zone of equiaxed crystals is substantially widened, the zone of columnar crystals is narrowed, and the degree of segregation is less dispersed. Figure 3.19

shows the macrostructure of slabs cast by the standard technology and with the introduction of a strip into the melt.

All of the above technological methods to some extent reduce the degree of the development of axial segregation by affecting the crystallization process. Nevertheless, along with the positive effects, each of the technologies considered has some significant drawbacks and does not guarantee the stability of the results.

3.4.2 Optimization of the Chemical Composition

The development of segregation processes to a certain extent depends on the steel chemical composition. Therefore, it is necessary to study the possibility of reducing the degree of the central segregation heterogeneity and its consequences by optimizing the chemical composition of the steel.

The most promising ways to upgrade low-alloy high-strength steels, especially the steels for large-diameter gas–oil pipes, are the reduction of carbon content and the replacement of the strengthening at the expense of pearlite by more advantageous strengthening mechanisms, primarily by grain refinement, formation of ferritic–bainitic microstructure, and precipitation hardening. In combination with other measures, such methods allow one to simultaneously increase the impact toughness, plasticity, resistance to brittle fracture, and weldability. The carbon content is a significant factor affecting the degree of segregation of chemical elements.

Decrease in carbon concentration leads to a decrease in the segregation of elements in the axial zone of the plates and a decrease in the intensity of development of segregation bands consisting of high-carbon structures of increased hardness, such as bainitic–martensitic bands (Fig. 3.20). Less pronounced axial segregation is the cause of the formation of bainitic or pearlitic segregation bands in the centerline

Fig. 3.20 Bainitic–martensitic segregation band in the axial plate zone (Belyj et al. 2005)

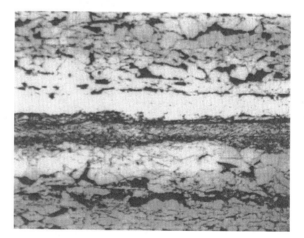

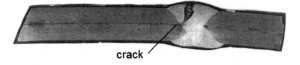

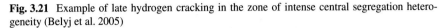

crack

Fig. 3.21 Example of late hydrogen cracking in the zone of intense central segregation heterogeneity (Belyj et al. 2005)

of the rolled product. The pronounced chemical segregation and increased hardness can cause late hydrogen-induced cracking upon the welding of plates (Fig. 3.21) or axial cracking and swelling upon the operation of the metal in a hydrogen sulfide-containing medium.

The effect of the mass fraction of carbon on the central chemical and structural segregation heterogeneity in continuously cast slabs from the 0.19%C-steel, 0.11%C-steel, 0.08%C-steel, and 0.03%C-steel low-alloy pipe steels was studied in (Belyj et al. 2005) (Table 3.4).

The steels differ substantially in carbon content, from 0.19%C to 0.03%C. Figure 3.22 shows the macrostructure of slabs from the test steels after etching at 60–80 °C in an aqueous solution of hydrochloric acid (50%). It is seen that, as the concentration of carbon in the steel decreases, the central segregation heterogeneity in slabs is substantially reduced. For example, the segregation heterogeneity is pronounced in the 0.19%C-steel slab and is not virtually detected in the 0.03%C-steel slab.

The estimation of axial chemical heterogeneity of slabs showed that, with decreasing carbon content in steel from 0.19 to 0.03% and sulfur content from 0.012 to 0.001%, the central segregation number decreases from 4 to 1 according to the OST 14-4-73 and from 5 to 1.5 according to the Mannesmann method (Table 3.5).

The slabs substantially differing in the central segregation depending on the carbon content were examined for the distribution of chemical elements, carbon, manganese, sulfur, phosphorus, niobium, and vanadium over their thickness (Fig. 3.23).

The maximum tendency to the segregation in the central zone of the slab is exhibited by sulfur (Fig. 3.24). For example, the sulfur segregation coefficient is 1.5 at 0.08% C and more than twice as large, 3.2, at 0.19% C. Phosphorus segregation coefficient also exhibits a noticeable sensitivity to the carbon content; it has increased from 1.19 at 0.03% C to 2.0 at 0.19% C.

A significant tendency to an increase in segregation with increasing carbon content in steel is exhibited by niobium. In the steel with 0.03% C, the segregation of niobium in the central zone of the slab was characterized by a coefficient of 1.23 at a niobium content of 0.086%, whereas in the steel with 0.08% C, K(Nb) was 1.7 at a substantially smaller niobium content of 0.029%. Vanadium compared with other elements is less prone to segregation within the carbon concentrations under consideration. The manganese segregation coefficient K(Mn) slightly increases from 1.05 to 1.25 with increasing carbon content from 0.03% to 0.19%. An increase in the content of carbon causes a rather moderate increase in the coefficient of its segregation, from 1.16 for

Table 3.4 Chemical composition of the steels with different carbon content

Steel	Elements (wt%)													C_{eq}	P_{cm}
	C	Mn	Si	S	P	Cr	Ni	Cu	Al	V	Nb	Ti		
0.19%C-steel	0.19	1.43	0.48	0.012	0.019	0.04	0.03	0.04	0.026	–	–	0.008	0.44	0.28
0.11%C-steel	0.11	1.69	0.29	0.004	0.015	0.03	0.03	0.02	0.034	0.099	0.029	0.015	0.41	0.22
0.08%C-steel	0.08	1.66	0.29	0.004	0.013	0.03	0.02	0.02	0.039	0.081	0.046	0.017	0.38	0.18
0.03%C-steel	0.03	1.49	0.16	0.001	0.013	0.27	0.16	0.25	0.024	–	0.086	0.011	0.38	0.14

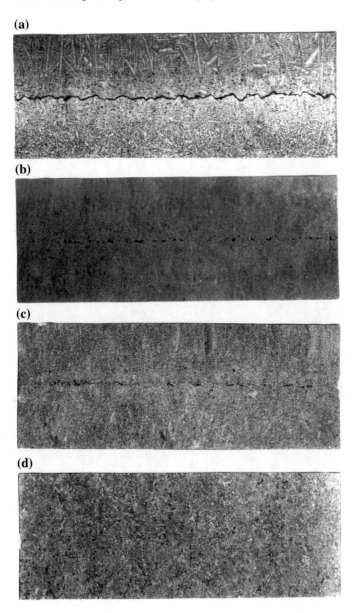

Fig. 3.22 Macrostructure of continuously cast slabs of steels with different carbon contents: **a** 0.19%C-steel; **b** 0.11%C-steel; **c** 0.08%C-steel; and **d** 0.03%C-steel

the steel with 0.03% C to 1.32 for the steel with 0.19% C. With increasing tendency to central segregation, the chemical elements in low-alloy steels are arranged in the following sequence: Mn→V→C→Nb→P→S.

Table 3.5 Central segregation heterogeneity of slabs from steels with different carbon contents

Steel	Content of elements (wt%)				Number of central segregation chemical heterogeneity	
	C	Mn	S	P	OST 14-4-73	Mannesmann
0.19%C-steel	0.19	1.43	0.012	0.019	4	5
0.11%C-steel	0.11	1.69	0.004	0.015	2	2.5
0.08%C-steel	0.08	1.66	0.004	0.013	1.5	2
0.03%C-steel	0.03	1.49	0.001	0.013	1	1.5

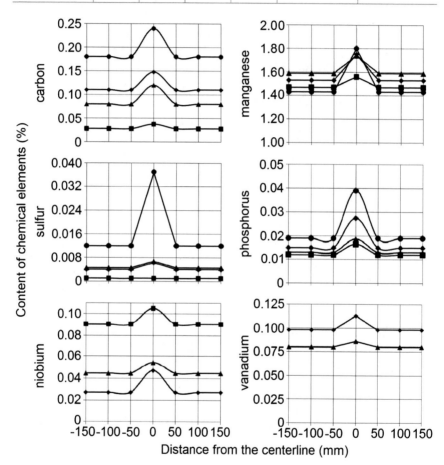

Fig. 3.23 Distribution of chemical elements over the slab thickness: filled circle—0.19%C-steel; filled diamond—0.11%C-steel; filled triangle—0.08%C-steel; filled square—0.03%C-steel

There is a distinct general regularity of decreasing tendency to segregation in the central zone of slabs with decreasing carbon content. The mechanism of such positive effect of reducing carbon concentration in the steels under consideration can

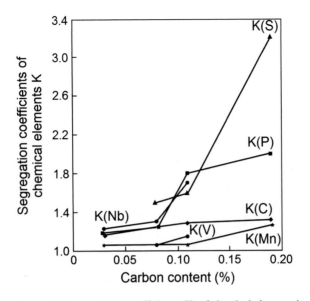

Fig. 3.24 Dependence of the segregation coefficients (K) of chemical elements in the axial zone of slabs on the total carbon content in steel

be explained with the help of the δ ferrite field of the iron–carbon phase diagram. In Fig. 3.25, vertical dashed lines in this region of the diagram correspond to the carbon contents of the investigated steels. It is seen that, as the carbon content in the steel decreases, the temperature range of the existence of δ ferrite expands and, correspondingly, the duration of the presence of metal in this region increases. As is known, the diffusivity of carbon atoms and impurities in δ ferrite is higher by several orders of magnitude than that is in austenite. This leads to a more homogeneous redistribution of the atoms of elements from their segregation zones (from interdendritic regions) throughout the volume. Primarily, this is characteristic of the steel with 0.03%C, the slabs of which exhibited the smallest central segregation heterogeneity.

The effect of low carbon concentrations ranging from 0.04 to 0.08% (at a manganese content of 1.25–1.35%) and the manganese content ranging from 0.65 to 1.35% (at a carbon content of 0.06%) on the central segregation of continuously cast slabs according to the Mannesmann evaluation method was studied in (Kholodnyi et al. 2016). The examination of the slabs from test steels showed a decrease in the average number of central segregation by about 1.0 (from 2.5 to ~1.5) with a decrease in the carbon concentration from 0.08 to 0.04% (Fig. 3.26a). As the manganese content was reduced from 1.35 to 0.65%, the average axial segregation decreases in class by more than 0.5, from >2 to 1.5 (Fig. 3.26b).

The change in the mass fractions of carbon, manganese, and phosphorus in the central zone of the slab from the steel containing (in %) 0.066 C, 0.26 Si, 1.26 Mn, 0.011 P, 0.001 S, 0.02 Al, 0.29 Cu, 0.04 V, and 0.03 Nb (Fig. 3.27) was analyzed in (Shabalov et al. 2017). The graphs of the distribution of chemical elements show that

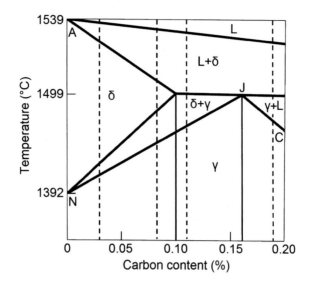

Fig. 3.25 Region of iron–carbon phase diagram with the δ ferrite field; the vertical dashed lines correspond to the carbon contents in the steels under study

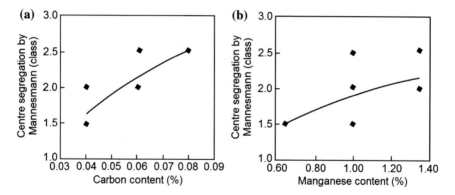

Fig. 3.26 Effect of the carbon content at Mn = 1.25–1.35% (a) and effect of manganese content at C = 0.06% (b) on the central segregation in continuously cast slabs according to Mannesmann estimation method (Kholodnyi et al. 2016)

the local maximum contents of the elements in the axial zone are substantially higher than their contents in the base metal. The carbon concentration increased from 0.066 to 0.110% (by a factor of 1.7), the manganese concentration increased from 1.26 to 1.55% (by a factor of 1.2), and the phosphorus concentration increased from 0.011 to 0.030% (by a factor of 2.7).

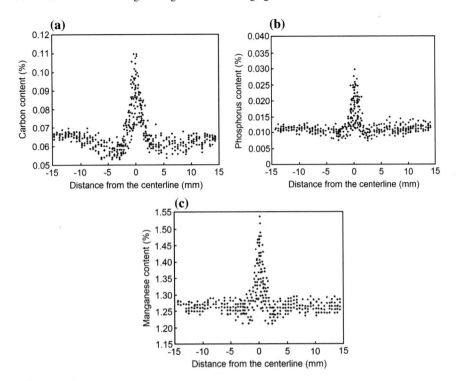

Fig. 3.27 Distribution of carbon (**a**), manganese (**b**), and phosphorus (**c**) in the axial zone of a continuously cast slab of steel containing (in %): 0.066 C; 0.26 Si; 1.26 Mn; 0.011 P; 0.001 S; 0.02 Al; 0.29 Cu; 0.04 V; 0.03 Nb (Shabalov et al. 2017)

Thus, the possibility of effective action on the central chemical and structural segregation heterogeneity is shown with the aim of reducing its degree by optimizing the chemical composition of steel. First of all, an adjustment is needed in the direction of reducing the content of carbon, manganese, phosphorus, sulfur, and niobium, which most strongly segregate in the axial zone of the continuously cast slab.

References

Belyj, A. P., Isayev, O. B., Matrosov, Yu I, et al. (2005). *Continuous casting blanks and rolling plates central segregation nonuniformity.* Moscow: Metallurgizdat.

Kholodnyi, A. A., Matrosov, Yu I, Matrosov, M Yu., & Sosin, S. V. (2016). Effect of carbon and manganese on low-carbon pipe steel hydrogen-induced cracking resistance. *Metallurgist, 60*(1), 54–60.

Shabalov, I. P., Matrosov, Yu I, Kholodnyi, A. A., et al. (2017). *Steel for gas and oil pipelines resistant to fracture in hydrogen sulphide-containing media.* Moscow: Metallurgizdat.

Chapter 4
Formation of the Structure and Properties of Plates from Pipe Steels upon Thermomechanical Processing with Accelerated Cooling

The plate and strip production from low-carbon low-alloy steels for oil and gas pipes and critical duty-welded structures is today one of the most innovative branches of ferrous metallurgy. Since the 1970s, there has been a continuous development of the chemical compositions of steels and rolling technologies in order to improve the properties combining high strength, toughness, cold resistance, corrosion resistance, and weldability of steels, as well as to reduce the cost of their production.

In addition to the chemical composition, the properties of plates are largely determined by its microstructure. A modern way to control the microstructure formation in steel is the combined process of controlled rolling and accelerated cooling. In this chapter, the physicometallurgical aspects of the microstructure formation and increase in mechanical properties of the plates manufactured by thermomechanical treatment by the scheme "controlled rolling followed by accelerated cooling" are considered. The investigated steels are intended for operation in hydrogen sulfide-containing media.

4.1   Effect of the Thermomechanical Processing Scheme

The plates are produced by various regimes of thermomechanical treatment or with the use of additional heat treatment. The effect of deformation-thermal treatment scheme on the microstructure and mechanical properties of plates 14–15 mm thick from hydrogen sulfide-resistant steel (0.06%C, 0.23%Si, 0.95%Mn, 0.15%Ni, 0.20%Cu, Ti + Nb + V $\leq$ 0.120%) was studied in (Matrosov et al. 2014a, Shabalov et al. 2017). The plates were produced in industrial conditions by the following regimes (Table 4.1):

© Springer Nature Switzerland AG 2019
I. Shabalov et al., *Pipeline Steels for Sour Service*, Topics in Mining, Metallurgy and Materials Engineering, https://doi.org/10.1007/978-3-030-00647-1_4

Table 4.1 Regimes of deformation-thermal treatment of the steel plates 14–15 mm thick

Treatment regime	T_{sr} (°C)	T_{fr} (°C)	T_{sc} (°C)	T_q (°C)	T_{fc} (°C)	V_c (°C/s)	T_t (°C)
HCR	910	830	–	–	–	–	–
LCR	810	745	–	–	–	–	–
CR + AC	900	840	800	–	505	20	–
LCR + Q + T	810	745	–	920	80	> 45	650

T_{sr} is the start rolling temperature in the finishing stand
T_{fr} is the finishing rolling temperature in the finishing stand
T_{sc} is the temperature start temperature accelerated cooling
T_q is the temperature of heating for quenching
T_{fc} is the finish temperature of the accelerated cooling
V_c is the cooling rate
T_t is the tempering temperature

- high-temperature controlled rolling with the end of deformation in the lower region of the γ field and cooling in air (HCR);
- low-temperature controlled rolling with the end of deformation in the ($\gamma + \alpha$) field and cooling in air (LCR);
- controlled rolling (CR) with the end of deformation in the lower part of the γ field and subsequent accelerated cooling (AC) from this region with interruption of AC in the field of bainitic transformation (CR + AC);
- low-temperature controlled rolling followed by quenching from special heating and tempering (LCR + Q + T).

The microstructure of the plates is shown in Fig. 4.1. The microstructure of the plates produced by the HCR scheme (Fig. 4.1a) consists of coarse-grained polygonal ferrite (PF) and lamellar pearlite (P). A decrease in the finish rolling temperature into the two-phase ($\gamma + \alpha$) field upon LCR led to the formation of a banded ferritic–pearlitic microstructure consisting of deformed grains of polygonal ferrite and lamellar pearlite (Fig. 4.1b). The microstructure of the steel after CR + AC was represented by a fine ferritic–bainitic mixture (Fig. 4.1c) consisting of the matrix of quasipolygonal ferrite (QPF) and an insignificant fraction of uniformly distributed regions of high-carbon upper bainite (UB) and cementite particles at the boundaries of ferrite grains. The use of LCR with subsequent quenching and tempering allowed a complete elimination of pearlite banding and the formation of a fine-grained ferritic–bainitic microstructure (Fig. 4.1d).

The tensile properties of the plates are shown in Fig. 4.2. The mechanical properties of the plates processed according to the LCR, CR + AC, and LCR + Q + T schemes were at comparable levels and corresponded to the X56 grade. The plates produced using the HCR technology had lower strength properties (X46 grade) and a $\sigma_{0.5}/\sigma_B$ ratio. At the same time, their plasticity was higher than that of plates treated by other regimes.

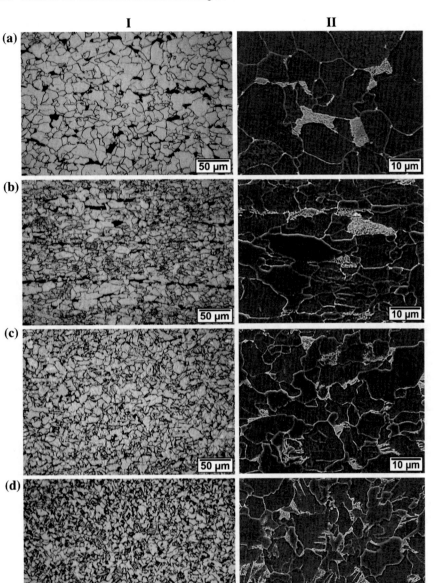

Fig. 4.1 Microstructure of the 0.06%C-0.23%Si-0.95%Mn-0.35%(Ni + Cu)-(Ti + Nb + V) ≤ 0.120% steel plates 14–15 mm thick manufactured by various deformation-thermal processing schemes, I—OM; II—SEM: **a** HCR; **b** LCR; **c** CR + AC; **d** LCR + Q + T

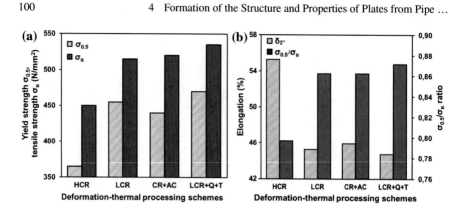

Fig. 4.2 Strength properties (**a**), elongation and $\sigma_{0.5}/\sigma_B$ ratio (**b**) of the 0.06%C-0.23%Si-0.95%Mn-0.35%(Ni + Cu)-(Ti + Nb + V) ≤ 0.120% steel plates 14–15 mm thick produced by various deformation-thermal treatment schemes

Of the listed schemes of deformation-thermal treatment of pipe steels, the most advanced technology is controlled rolling with subsequent regulated cooling. According to the terminology accepted in the world literature, this technology is designated Thermomechanical Controlled Processing (TMCP). This definition includes the combination of Controlled Rolling (CR) and Accelerated Cooling (AC).

Post-deformation accelerated cooling allows one to simultaneously increase the strength and cold resistance of plates by structure refinement and substitution of the ferritic–pearlitic microstructure for the ferritic–bainitic one. The use of the positive effect of accelerated cooling on the properties of plate metal required the development of new compositions of low-alloy steels to make full use of the advantages of this technology. The formation of the optimum microstructure of the plates provides the combination of high strength, plasticity, the resistance to brittle and ductile fracture at low climatic temperatures as well as the resistance to cracking in hydrogen sulfide-containing media, weldability, and other special properties.

Figure 4.3 shows the scheme of the CR and CR + AC technologies and the microstructure state of rolled products at each stage of thermomechanical processing

Upon the manufacture of plates by the CR technology, rolling usually is finished in a two-phase ($\gamma + \alpha$) field at temperatures of about 680–780 °C, and then the plates are cooled in air. This leads to the formation of a deformed ferritic–pearlitic microstructure. Upon processing by the CR + AC scheme, the deformation ends in a single-phase γ field (rarely, in the upper part of the ($\gamma + \alpha$) field) and then the plates are cooled at a rate usually ranging between 10 and 30 °C/s to a predetermined temperature. This technology leads to the formation of a finer ferritic–bainitic microstructure and reduces the structure banding.

Application of accelerated cooling after controlled rolling allows one to reduce content of alloying and microalloying elements in the steel. Due to more economic alloying, the carbon equivalent is reduced for all grades. This is especially important for thicker plates, and also for improving weldability in factory and field conditions. An important role is played by the possibility to reduce the wear of equipment and to

Fig. 4.3 Schematic illustration of the CR and CR + AC processes: I—field of complete recrystallization of austenite; II—field of partial recrystallization of austenite; III—field of the absence of austenite recrystallization; IV—field of the γ→α transformation (Matrosov et al. 2014)

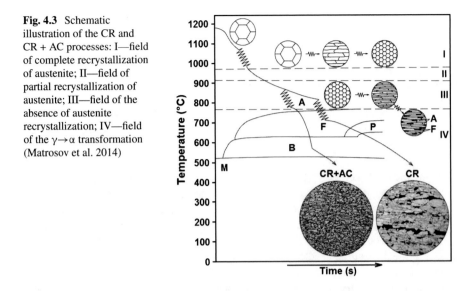

increase the productivity of the hot rolling mill by elevating the finishing deformation temperature. Due to such advantages, the CR + AC technology became widely used in modern metallurgical enterprises manufacturing rolled plates for electrically welded pipes and other critical metal structures.

4.2   Post-deformation Cooling

At present, most technological lines of plate reversible rolling mills are equipped with controlled accelerated cooling units. The principle of their operation is based on the regulated supply of water to the hot metal for the purpose of heat removal, i.e., cooling of rolled stock. The main parameters of post-deformation-controlled cooling are as follows: the temperature of the start and interruption of cooling, cooling rate, and uniformity of cooling over the area and thickness of rolled products. The versions of accelerated cooling can be divided into three types:

- accelerated cooling, at which the surface temperature of the rolled product does not decrease below the martensitic point (M_s), with subsequent cooling in air;
- direct quenching, at which cooling of rolled products occurs at a high rate to a temperature below M_s, and then the product is subjected to tempering;
- temper quenching including rapid cooling, at which the surface temperature of the rolled product falls below M_s and then increases above M_s due to the heat of the inner layers of the metal.

In accordance with the type of water supply to the hot metal, the methods of accelerated cooling can be divided into cooling with a water curtain or laminar jets, cooling with an air–water mixture (Adco), and cooling with a water pillow (Mulpic). Laminar cooling devices are characterized by a low flow rate of a water jet providing a high cooling capacity. Water–air units (mist cooling) provide cooling by a jet of finely dispersed water droplets injected under pressure and sprayed with air. The industrial accelerated cooling units can use either one cooling type or a combination of cooling types.

There are two methods of accelerated cooling in the mill flow, simultaneous and sequential cooling, depending on the transport scheme of the cooled plate. In the former case, the cooling water is fed directly to the entire surface of the rolled plate, with a reciprocating movement of the plate in the cooling zone for uniform cooling. This scheme has several drawbacks since the cooling zone should be longer than the rolled plate, and, in addition, water can retain on the surface after cutting-off its feed and cause local overcooling. For these reasons, the cooling systems feeding water to the plate moving through a cooling zone, which is substantially shorter than that is upon simultaneous cooling of the entire plate, have become more widespread.

Great attention at the accelerated cooling is paid to its uniformity over the rolled product area and to the elimination of the plate buckling. The industrial equipment provides for various measures aimed at increasing the uniformity of cooling, including masking the plate edges, the optimal distribution of water flow over the plate area. To control the cooling of the front and back ends of the plate, they are "masked" by interrupting the water supply for the time of the passage of each end. Buckling of the plate is also caused by the difference in temperatures of its upper and lower surfaces. For this reason, methods are used to balance the cooling capacities of the upper and lower nozzle groups.

Figure 4.4 shows an SMS-Demag controlled cooling unit (CCU) of laminar type installed at a distance of ~45 m behind the finishing stand in the flow of a 3600 plate mill. The total CCU length is 35,100 mm including a cooling section of 25,600 mm, and the cooling zone width is 4100 mm. Roller table consists of 43 solid cooled rollers of 400 mm in diameter and 4200 mm in barrel length. The linear speed of the plate movement along the roller table reaches 2.5 m/s upon cooling and up to 3.5 m/s upon transportation. The cooled plates are limited by 5–50 mm in thickness, 1600–4050 mm in width, up to 40,000 mm in length, and up to 23.5 tonnes in weight.

The cooling system includes 15 upper and 15 lower collectors. The rolled plates are cooled from above with the help of U-shaped pipes creating laminar jet stream of water and from below by quasilaminar water flow created by horizontal irrigation pipes.

To remove water from the rolled plates, side nozzles are installed at the front, middle, and end of the device in the transverse direction. Depending on the width of the cooled plate, the width of the cooling zone is controlled by the edge shielding system. After cooling, the strip is blown by compressed air.

The cooling rate of the rolled plate depends on the initial and end cooling temperatures, the plate thickness, the speed of plate movement on the roller table, the water temperature, and the intensity of the water flow. Table 4.2 shows the guaranteed maximum cooling rate as a function of the plate thickness. The cooling system is automatically controlled using a mathematical model that calculates the unit opera-

Fig. 4.4 SMS-Demag controlled cooling unit of the laminar type in the 3600 plate mill flow

Table 4.2 Maximum cooling rate of plates in controlled cooling unit

Plate thickness (mm)	Maximum cooling rate (°C/s)
10	45
20	24
25	20
30	17
40	12

Start cooling temperature is 800 °C
Finish cooling temperature is 500 °C
Temperature of cooling water is 25 °C

tion parameters on the basis on the input data. To provide a predetermined cooling rate, the mathematical model determines the required water flow and the speed of plate movement along the roller table with allowance for the actual temperature parameters of the rolled product.

A hot mangle is installed behind the accelerated controlled cooling unit in the mill line (Fig. 4.5).

4.2.1 Structure Transformations upon Cooling

For the development of new technologies, leading research centers and metallurgical companies tend to make the most use of the results of laboratory research, which allows one to substantially reduce the costs of the implementation of indus-

Fig. 4.5 Hot mangle behind the controlled cooling unit in the 3600 plate mill flow

trial production (Efron 2012; Ringinen et al. 2014). The processes occurring upon heating, hot deformation, and cooling and their effect on the microstructure and properties of steel are studied by simulating on laboratory equipment. It is known that post-deformation cooling strongly affects the microstructure and properties of rolled metal (Bufalini et al. 1983, Marosov et al. 2014c). In this association, great attention is paid to the study of the specific features of the structure transformations occurring in steel upon cooling after hot plastic deformation. The method of constructing thermokinetic diagrams (TKD) is widely used for the study of the austenite transformations (Efron 2012; Ringinen et al. 2014; Salganik et al. 2016). Below we consider the results on the kinetics of hot-deformed austenite decomposition upon continuous cooling in low-carbon low-alloy pipe steels of various alloying systems used for manufacturing large-diameter gas–oil pipes in cold-resistant and hydrogen sulfide-resistant designs (Matrosov et al. 2014a, Efimov et al. 2017).

The thermokinetic diagrams were plotted from the cooling curves recorded with a BÄHR-805 fast-acting deformation dilatometer using cylindrical specimens of 5 mm in diameter and 10 mm in length from industrially manufactured plates. The temperature conditions of heating and deformation of the specimens were assigned in such a way that they corresponded most closely to the actual regimes of high-temperature controlled rolling with the end of deformation in the austenite region. The treatment regimes in the dilatometer simulated the controlled rolling in the mill by providing for heating at a rate of 1.5 °C/s to a temperature of 1150 °C with subsequent holding for 3 min, deformation by compression to 15–20% upon slow cooling at temperatures of 1050, 850 and 820 °C, further cooling to a temperature of 100 °C at a rate ranging from 0.5 to 100 °C/s.

The following main parameters of the $\gamma \rightarrow \alpha$ transformation occurring at different cooling rates were determined by the analysis of the dilatometric cooling curves and the study of the specimens: the positions and extensions of the ferrite, pearlite, bainite, and martensite formation regions, the volume fractions of the structure constituents, and the steel hardness. The obtained results were plotted on the graph with the "temperature-cooling rate" coordinates.

Table 4.3 shows the chemical compositions of the low-carbon pipe steels used to produce X60–X70 grade plates satisfying high requirements for cold resistance. The steels were characterized by different C contents and additions of alloying (Mn, Cr, Ni, Cu, Mo) and microalloying (Nb, V) elements. For comparison, the conventional construction steel of the C–Si–Mn alloying system was studied.

The thermokinetic diagrams of the decomposition of hot-deformed austenite upon continuous cooling of test steels are shown in Fig. 4.6.

The thermokinetic diagram of the steel 1 (C↑ + Mn↓) of the C–Si–Mn alloying system exhibits a wide temperature-rate region of the ferritic–pearlitic structure formation at cooling rates of ≤20 °C/s. The ferritic–pearlitic–bainitic microstructure was detected at a cooling rate of 30 °C/s. The ferritic–bainitic structure was formed at cooling rates ranging from >30 to 100 °C/s.

A reduction in the carbon content to 0.09%, a substantial increase in the mass fraction of manganese to 1.66%, and the introduction of microalloying elements into the steel 2 (C + Mn↑ + Ti + Nb + V) shifted the pearlitic transformation range to lower cooling rates compared to those of the steel 1 (C↑ + Mn↓), and no pearlite was found in the steel 2 (C + Mn↑ + Ti + Nb + V) at cooling rates of ≥10 °C/s. At the same time, bainite formation was detected in the microstructure of the steel 2 (C + Mn↑ + Ti + Nb + V) at a cooling rate of 2 °C/s, and martensite was present at rates of ≥ 30 °C/s. Such change in the chemical composition of the steel affected the temperature of the onset of the $\gamma \rightarrow \alpha$ transformation: at a cooling rate of 1 °C/s, the Ar_3 temperature was 820 °C for the steel 1 (C↑ + Mn↓) and decreased to 750 °C for the steel 2 (C + Mn↑ + Ti + Nb + V) steel.

Table 4.3 Chemical composition of the steels

Steel	Elements (wt%)									
	C	Si	Mn	Cr	Ni	Cu	Mo	Ti	Nb	V
Steel 1 (C↑ + Mn↓)	0.16	0.21	0.58	–	–	–	–	–	–	–
Steel 2 (C + Mn↑ + Ti + Nb + V)	0.09	0.27	1.66	–	–	–	–	0.017	0.050	0.076
Steel 3 (C↓ + Mn + Mo + Ti + Nb)	0.05	0.20	1.43	–	–	–	0.16	0.016	0.050	–
Steel 4 (C↓ + Mn + Cr + Ni + Cu + Ti + Nb)	0.05	0.20	1.42	0.27	0.21	0.27	–	0.014	0.080	–

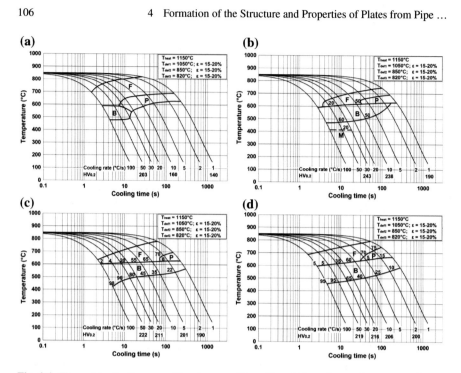

Fig. 4.6 Thermokinetic diagrams of the decomposition of hot-deformed austenite upon continuous cooling of steels: **a** steel 1 (C↑ + Mn↓), **b** steel 2 (C + Mn↑ + Ti + Nb + V), **c** steel 3 (C↓ + Mn + Mo + Ti + Nb), **d** steel 4 (C↓ + Mn + Cr + Ni + Cu + Ti + Nb)

A decrease in the carbon content to 0.05% and additional alloying with 0.16% molybdenum [the steel 3 (C↓ + Mn + Mo + Ti + Nb) compared with the steel 2 (C + Mn↑ + Ti + Nb + V)] led to the suppression of pearlitic transformation at cooling rates of <5 °C/s. The austenite decomposition in this steel at all cooling rates used in the study was finished by the transformation in the intermediate (bainitic) field. An increase in the cooling rate to 100 °C/s did not reveal any tendency to the formation of martensite. This is explained, first of all, by low carbon content. At the same time, due to a decrease in the quantity of polygonal ferrite, the fraction of bainite in the steel structure increased.

The effect of simultaneous additions of chromium (0.27%), nickel (0.21%), and copper (0.27%) on the kinetics of the austenite decomposition in the steel 4 (C↓ + Mn + Cr + Ni + Cu + Ti + Nb) is similar to that of the alloying of the steel 3 (C↓ + Mn + Mo + Ti + Nb) with 0.16% molybdenum. However, the pearlitic transformation in the steel 4 (C↓ + Mn + Cr + Ni + Cu + Ti + Nb) steel was observed at cooling rates of 5 °C/s and lower.

In general, it can be noted that alloying with molybdenum or with joint additions of nickel, chromium, and copper positively affects the structure of the steel, since it suppresses the formation of pearlite and promotes the formation of the ferritic–bainitic structure over a wide range of cooling rates.

Table 4.4 Chemical composition of the steels

Steel	Elements (wt%)							
	C	Mn	Si	Cr	Ni	Cu	Mo	Ti + Nb + V
Steel 1 (Cr)	0.05–0.07	0.90–0.95	0.20–0.23	0.25	–	–	–	≤0.120
Steel 2 (Cr + Ni + Cu)					0.25	0.20–0.25	–	
Steel 3 (Cr + Ni + Cu + Mn↑)		1.25					–	
Steel 4 (Cr + Ni + Cu + Mo)		0.93					0.15	

Low-alloy pipe steels used for the manufacture of rolled products cracking resistant in H_2S-containing media are characterized by a decreased manganese content, which usually does not exceed 1.25% (Kholodnyi et al. 2016). To compensate for the reduced manganese content, alloying with Cr, Ni, Cu, and Mo is often used. This specific feature of the chemical composition of the H_2S-resistant steels substantially affects the structure transformation of austenite upon cooling. In this association, it was of interest to study the kinetics of phase transformations, microstructure, and properties of H_2S-resistant steels of various alloying systems. The specimens for the study were taken from industrially manufactured plates of chemical compositions given in Table 4.4.

It is seen from Table 4.4 that the test steels, except the basic alloying elements such as (0.05–0.07%) C—(0.90–0.95%) Mn—(0.20–0.23%) Si—(Ti + Nb + V) ≤ 0.120%, contained various combinations of additional alloying elements (up to 0.25% each in various combinations, up to a total of 0.90%): the steel 1 (Cr) contained Cr, the steel 2 (Cr + Ni + Cu) contained Cr + Ni + Cu, the steel 3 (Cr + Ni + Cu + Mn↑) contained Cr + Ni + Cu and an increased amount of Mn (1.25%), and the steel 4 (Cr + Ni + Cu + Mo) contained Cr + Ni + Cu + Mo.

Figure 4.7 shows the thermokinetic diagrams indicating the presence of the regions of ferritic, pearlitic, and bainitic transformations in the investigated range of cooling rates of the test steels. An increase in the degree of alloying leads to a narrowing of the temperature interval for the formation of ferrite and to the expansion of the bainitic transformation region. The TKD of the steel 1 (Cr) (Fig. 4.7a) exhibits a wide region of the pearlitic transformation, which can occur at a cooling rate of up to ~25 °C/s, while additional introduction of the alloying elements intensely suppresses the pearlitic transformation, which was absent upon cooling at a rate of ≥10 °C/s for the steel 2 (Cr + Ni + Cu) and at a rate of ≥5 °C/s for the steel 3 (Cr + Ni + Cu + Mn↑) and steel 4 (Cr + Ni + Cu + Mo) (Fig. 4.7b–d).

The formation of the ferritic–bainitic microstructure is possible upon cooling in the steel 1 (Cr) at a rate of ≥30 °C/s, in the steel 2 (Cr + Ni + Cu) at a rate of ≥ 10 °C/s, and in the steel 3 (Cr + Ni + Cu + Mn↑) and steel 4 (Cr + Ni + Cu + Mo) at a rate of ≥5 °C/s. In this case, the bainitic transformation was recorded upon cooling at a rate of 5 °C/s for the steel 1 (Cr), steel 2 (Cr + Ni + Cu), and steel 3 (Cr + Ni + Cu + Mn↑) steels, and at a rate of 2 °C/s for the steel 4 (Cr + Ni + Cu + Mo).

Figure 4.8 shows the microstructure of the steel 2 (Cr + Ni + Cu) cooled at different rates: 1, 5, 20, and 50 °C/s. It is seen that, after cooling at a rate of 1 °C/s,

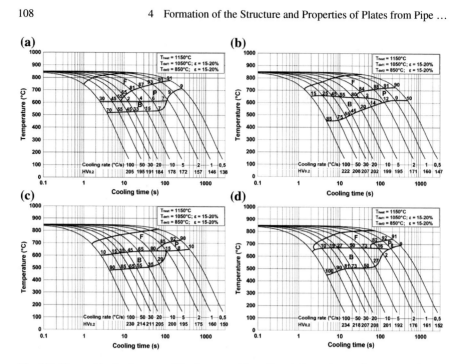

Fig. 4.7 Thermokinetic diagrams of the decomposition of hot-deformed austenite upon continuous cooling of H_2S-resistant pipe steels of various alloying systems: **a** steel 1 (Cr); **b** steel 2 (Cr + Ni + Cu); **c** steel 3 (Cr + Ni + Cu + Mn↑); **d** steel 4 (Cr + Ni + Cu + Mo)

the microstructure is ferritic–pearlitic. An increase in the cooling rate to 5 °C/s led to the formation of bainite and a decrease in the fraction of pearlite and ferrite. A further increase in the cooling rate increases the fraction of fine acicular products of the intermediate transformation with a simultaneous decrease in the fraction of ferrite.

Figure 4.9 shows the effect of cooling rate on the fraction of the ferrite and bainite microstructure constituents and the hardness of the test steels. It is seen that, with increasing cooling rate, the fraction of ferrite decreases, and the fraction of bainite simultaneously increases. At similar cooling rates, the volume fraction of bainite is higher and the volume fraction of ferrite is smaller in the steels in the following order with increasing degree of alloying: Cr→Cr + Ni + Cu→Cr + Ni + Cu + Mn↑→Cr + Ni + Cu + Mo.

The formation of fully bainitic microstructure took place in the Cr + Ni + Cu + Mo steel upon cooling at a rate of 100 °C/s. In the steel with the addition of only Cr, the fraction of bainite after cooling at a rate of 100 °C/s was 70%.

The steel 1 (Cr) exhibited the lowest hardness, and the highest hardness was observed for the steel 4 (Cr + Ni + Cu + Mo). The intense growth of hardness occurs as the cooling rate increases from 0.5 to 10 °C/s, which corresponds to a decrease in the ferrite grain size and the beginning of the formation of intermediate transformation products in the steels under study.

(a)   91%F/9%P   **(b)**   84%F/2%P/14%B

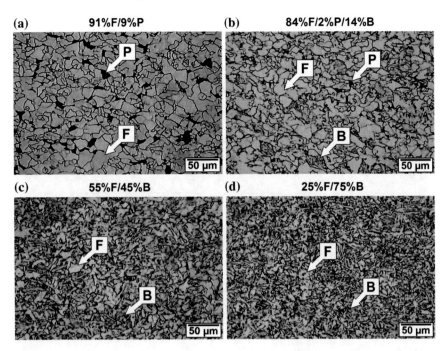

(c)   55%F/45%B   **(d)**   25%F/75%B

Fig. 4.8 Microstructure of the steel 2 (Cr + Ni + Cu) after cooling at different rates, OM: **a** 1 °C/s; **b** 5 °C/s; **c** 20 °C/s; **d** 50 °C/s

The microstructures of the steel 1 (Cr) and steel 4 (Cr + Ni + Cu + Mo) steels cooled at a rate of 30 °C/s are compared in Fig. 4.10. The ferrite and bainite contents in the steel 1 (Cr) were 55%F and 45%B, whereas in the steel 4 (Cr + Ni + Cu + Mo) additionally alloyed with Ni + Cu + Mo, the ratio between the ferrite and bainite fractions was 19%F/81%B.

Figure 4.11 shows the effect of the additions of alloying elements on the yield strength and the tensile strength of industrial plates 18–20 mm thick from the steels manufactured using the same regimes of controlled rolling ($T_{fr} = Ar_3 + (30–50)$ °C) and accelerated cooling ($T_{sc} = Ar_3 + (10–30)$ °C, $T_{fc} = 500–550$ °C, $V_c = 20–30$ °C/s). It is seen that, simultaneously with the increase in microhardness, the strength characteristics of the plates increased on average by 75–80 N/mm^2 with increasing degree of alloying from Cr to Cr + Ni + Cu + Mo. The strength levels of the plates from the steel 1 (Cr) and steel 2 (Cr + Ni + Cu) corresponded to the X52 and X60 grades, respectively, while the plates from the steel 3 (Cr + Ni + Cu + Mn↑) and steel 4 (Cr + Ni + Cu + Mo) corresponded to the X65 grade.

The production technology of low-carbon low-alloy pipe steels provides maximum refinement and strengthening of austenite grains upon the roughing and finishing stages of controlled rolling. The fine flattened austenite grains with a large specific surface contribute to an increase in temperature and acceleration of the normal diffusion-controlled $\gamma \rightarrow \alpha$ transformation. In the case of rapid diffusion of carbon

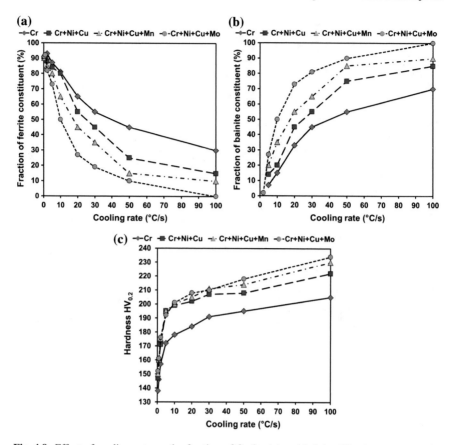

Fig. 4.9 Effect of cooling rate on the fraction of ferrite (**a**) and bainite (**b**) microstructure constituents and hardness (**c**) of the steels basically containing (0.05–0.07%) C—(0.90–0.95%) Mn—(Ti + Nb + V) ≤ 0.120% of various alloying systems

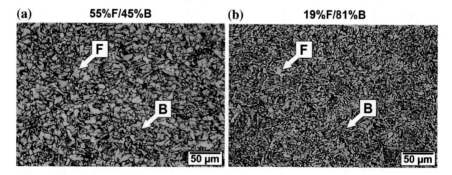

Fig. 4.10 Microstructure of the test steels after cooling at a rate of 30 °C/s, OM: **a** steel 1 (Cr); **b** steel 4 (Cr + Ni + Cu + Mo)

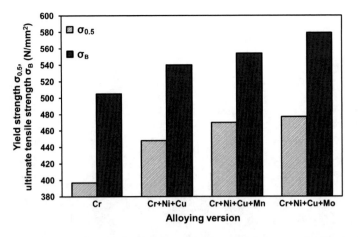

Fig. 4.11 Strength properties of industrial plates 18–20 mm thick from steels of basic chemical composition (0.05–0.07%) C—(0.90–0.95%) Mn—(0.20–0.23%) Si—(Ti + Nb + V) ≤ 0.120% of various alloying systems after controlled rolling with accelerated cooling

(including grain-boundary diffusion), the polyhedral ferrite grains grow by the normal (recrystallization) mechanism, at which carbon outflow causes the formation of local regions enriched with carbon. The transformation products become separated into high-carbon structure constituents (pearlite, high-carbon bainite, martensite, austenite) and a low-carbon phase represented by various forms of ferrite. The volume fraction of the low-carbon α phase is much higher than the volume fraction of the high-carbon structure constituents. In this case, the type and morphology of structure constituents substantially depend not only on chemical composition, but also on the temperature-deformation regimes of controlled rolling and the temperature-rate conditions of post-deformation cooling (Matrosov et al. 2012). Figure 4.12 shows the plate microstructures consisting of low-carbon ferritic matrix and high-carbon structures such as ferritic–pearlitic microstructure formed after controlled rolling with air cooling and ferritic–bainitic microstructure formed after controlled rolling and accelerated cooling.

The specific features of the chemical composition and processing technology of the pipe steels cause certain difficulties in the identification and classification of the microstructures compared with those of high- and medium-carbon steels. Low carbon content (usually <0.08%C), intense grain refinement, and accelerated cooling upon TMCP lead to the final structure consisting mainly of intermediate transformation products with indistinct morphological characteristics. Despite the fact that such structures are classified as bainitic, the ferrite grains (regions) are free from plate carbide precipitates typical of the upper and lower bainite structures in medium-carbon steels. After TMCP, the structure of low-carbon steels is a mixture of different types of ferrite matrix and hard phases sequentially precipitating at different temperatures upon continuous cooling. It is extremely difficult to characterize such mixture of products by one term. To identify and classify microstructures, the authors of

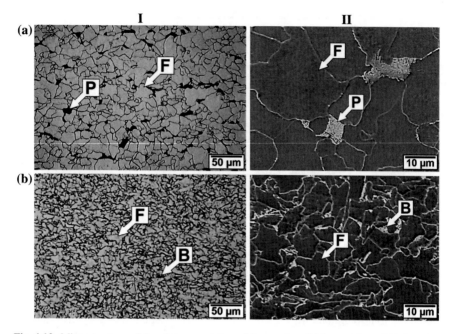

Fig. 4.12 Microstructure of the plates consisting of low-carbon ferrite matrix and high-carbon structures, I—OM; II—SEM: **a** ferritic–pearlitic; **b** ferritic–bainitic

(Matrosov et al. 2012) proposed to use a fairly simple method based on the approach developed by the ISIJ Bainite Committee with some refinements. No introduction of new names for numerous varieties of complex structures obtained in steels of different compositions at different TMCP regimes is required by the proposed method. The classification is based on separate description: of the morphology of the ferrite matrix, which occupies most of the volume; of the type and morphology of "island" structure constituents enriched with carbon (their content is usually <10%); and of the morphology and composition of inclusions or interlayers in "regular" structures. The "island" structures mean the regions enriched with carbon and differing in structure from the low-carbon ferrite matrix. The spacing between such islands is larger by several folds than the size of the matrix structure constituent. At the same time, there are no orientation relationships between the high-carbon structure constituent and the neighboring regions of the low-carbon structure. Such regions can be considered as isolated structure constituents, for example, pearlite in ferritic–pearlitic structure. The type and morphology of the "island" structures are determined by the local carbon concentrations, which in turn depend on the chemical composition of the steel and the deformation-thermal processing parameters.

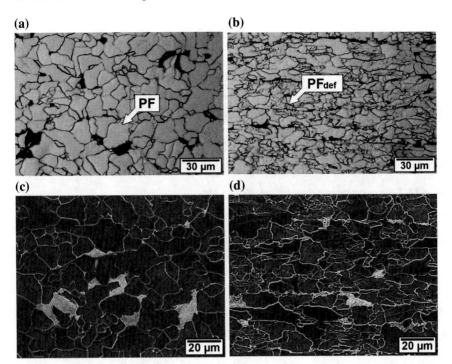

Fig. 4.13 Polygonal ferrite (PF): **a, b** OM; **c, d** SEM

The low-carbon α phase in the H$_2$S-resistant pipe steels can be represented by the following microstructures in order of decreasing their transformation temperature:

- polygonal (polyhedral, equiaxed) ferrite (PF) is formed upon the transformation of austenite by a normal diffusion-controlled mechanism at temperatures above 650 °C (Fig. 4.13). Ferrite is characterized by equiaxed grains and a low dislocation density. The grain boundaries are mainly high-angle. In the case of plastic deformation of ferrite, the grains are elongated along the rolling direction and have a higher dislocation density (Fig. 4.13b, d);

- quasipolygonal (quasipolyhedral) ferrite (QPF) is formed under conditions of the inhibition of diffusion processes in austenite (sufficiently high cooling rate and increased level of alloying), which in this case have no time for transformation into polygonal ferrite at higher temperatures. Compared to PF, QPF has a smaller crystal size, substructure consisting of weakly misoriented subgrains, a higher dislocation density, and curved crystal boundaries (Fig. 4.14);

- granular and acicular bainitic ferrites (GBP and ABF, respectively) are formed by the bainitic (diffusion-shear) mechanism upon cooling at high rates to a temperature below 600 °C under conditions of virtually complete suppression of diffusion-controlled processes. The designation of bainitic ferrite is based on its morphology;

(a)   **(b)**

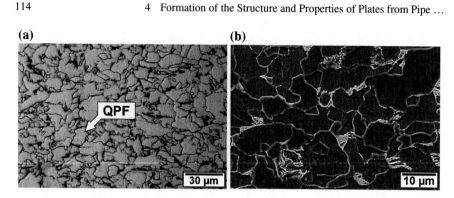

Fig. 4.14 Quasipolygonal ferrite (QPF): **a** OM; **b** SEM

- low-carbon lath martensite (M) is formed at high rates of cooling to temperatures below Ms by the mechanism of martensitic transformation, when carbon has no time to precipitate from the solid solution.

In accordance with decreasing transformation temperature, the island high-carbon structures are as follows:

- lamellar pearlite (P) is formed by the diffusion-controlled mechanism from the austenite regions enriched with carbon and alloying elements, mainly at low cooling rates (usually <5 °C/s) and at temperatures above 600 °C. Pearlite is a eutectoid mixture of α phase and thin parallel cementite plates (Fig. 4.15);
- degenerate pearlite (DP) is formed if the time is insufficient for the occurrence of diffusion-controlled processes. In this case, the regularity of the arrangement of the cementite plates is disrupted, and the colonies are refined (Fig. 4.16);
- high-carbon upper bainite (UB) is formed from austenite regions enriched with carbon at temperatures below than those of the pearlite formation. The UB structure is a mixture of the α-phase and cementite plates (lens), which are finer and more misoriented than those are in pearlite (Fig. 4.17);
- MA constituent (high-carbon twinned martensite + retained austenite) is formed upon supercooling of microscopic volumes of austenite with a high concentration of carbon and alloying elements. The structure of the MA constituent contains a certain amount of retained austenite, the fraction of which depends on the temperature corresponding to the point of the maximum possible quantity of martensite in the local region of this structure constituent. The regions of the MA constituent are usually up to 5 μm in size (Fig. 4.18).

Cementite (C) in the plates treated by various regimes of thermomechanical treatment is found in the form of precipitates at the ferrite grain boundaries (Fig. 4.19).

High-carbon island structures are more often present in the environment of polyhedral or quasipolygonal ferrite grains. This is explained by the need for a significant redistribution of carbon upon the γ→α transformation.

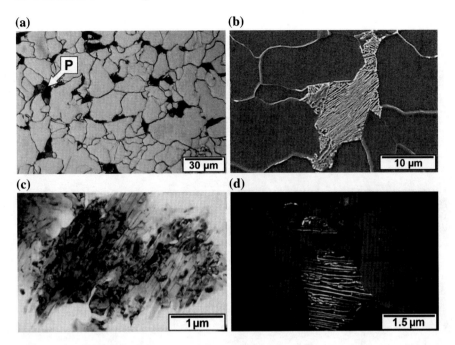

Fig. 4.15 Lamellar pearlite (P): **a** OM; **b** SEM; **c** TEM (light field); **d** TEM (dark-field in the cementite reflection)

4.2.2 Effect of the Accelerated Cooling Regimes

This section considers the effects of accelerated cooling parameters such as the start and finish temperatures of the accelerated cooling (T_{sc} and T_{fc}, respectively) and the cooling rate (V_c) on the microstructure and mechanical properties of the plates (Matrosov et al. 2014c).

Start Temperature of the Accelerated Cooling Upon processing by the CR + AC technology, the start temperature of the accelerated cooling of the plates depends on the temperature of the finish of rolling in the finishing stand and the thickness of the rolled plate at a constant rate of its movement on the roller table from the stand to the controlled cooling unit.

In the production conditions in plate rolling mills, the following versions of the T_{fr} and T_{sc} regimes are possible relative to the start temperature of the $\gamma \rightarrow \alpha$ transformation (critical point Ar_3):

- the end of deformation and the start of accelerated cooling from the single-phase γ field: $T_{fr} > Ar_3$ and $T_{sc} > Ar_3$;
- the end of deformation in the single-phase γ field and the start of accelerated cooling from the two-phase ($\gamma + \alpha$) field: $T_{fr} > Ar_3$ and $T_{sc} < Ar_3$;

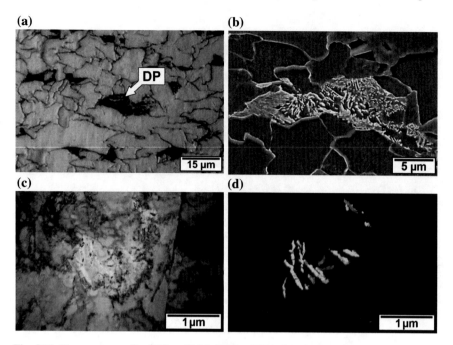

Fig. 4.16 Degenerate pearlite (DP): **a** OM; **b** SEM; **c** TEM (light field); **d** TEM (dark-field image in the cementite reflection)

- the end of deformation and the start of accelerated cooling from the two-phase $(\gamma + \alpha)$ field: $T_{fr} < Ar_3$ and $T_{sc} < Ar_3$.

The effect of the T_{fr} and T_{sc} regimes on the microstructure and properties was studied on plates 18 mm thick from the 0.05%C-1.25%Mn-0.70%(Cr + Ni + Cu)-(Ti + Nb + V)-steel manufactured according to the regimes given in Table 4.5. The T_{fr} and T_{sc} regimes of the experimental plates were changed in the study, but the temperatures of the finish of the accelerated cooling and the cooling rate were the same.

The change in the finishing rolling temperatures and the start temperature of accelerated cooling in the used intervals did not substantially affect the mechanical properties of the plates.

The effect of the finishing rolling temperature and the start temperature of accelerated cooling on the microstructure of the plates is shown in Fig. 4.20. After accelerated cooling from the single-phase austenite field (regime I), the microstructure was a fine ferritic–bainitic mixture consisting predominantly of quasipolygonal ferrite and islands of high-carbon bainite (Fig. 4.20a).

After the treatment by regime III (the end of deformation and the start of accelerated cooling from the two-phase $(\gamma + \alpha)$ field), the microstructure of the base metal consists of the polygonal ferrite matrix and insignificant fraction of QPF. In this case, the fractions and sizes of high-carbon structures in the form of degenerate pearlite

(a)   **(b)**

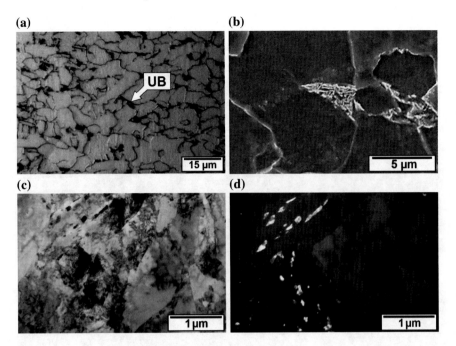

(c)   **(d)**

Fig. 4.17 High-carbon upper bainite (UB): **a** OM; **b** SEM; **c** TEM (light field); **d** TEM (dark-field image in the cementite reflection)

(DR) and high-carbon bainite increase relative to those of the state after processing by regime I, (Fig. 4.20b). In view of a sufficiently high finishing rolling temperature, there is no substantial deformation of the ferrite grains.

Figure 4.21 shows the dependences of the strength properties of the 0.05%C-1.43%Mn-0.16%Mo-0.016%Ti-0.050%Nb-steel plates on the start temperature of the accelerated cooling. The dependences exhibit maxima at temperatures of 775–795 °C. The start temperature of the $\gamma \rightarrow \alpha$ transformation for the steel was about 775 °C. A decrease in T_{sc} below the Ar$_3$ temperature or a substantial elevation of T_{sc} above Ar$_3$ (in this case, by 60 °C) led to a decrease in strength properties. At the same time, the relative elongation, Charpy impact energy, and shear area of the DWTT fracture surface did not noticeably change: $\delta_{2''} = 29$–31%; KV^{-20} = 266–300 J, DWTT^{-20} = 100%.

Finish Temperature of Accelerated Cooling The effect of the finish temperature of accelerated cooling on the plate microstructure and properties has been studied on the 0.07%C-1.33%Mn-0.70%(Cr + Ni + Cu)-(Ti + Nb + V)-steel plates 20 mm thick, which were manufactured with the finish of accelerated cooling at various temperatures in a range of 430–650 °C. In this case, $T_{sc} = $ Ar$_3$ + (0–20) °C, $V_c = $ 16–19 °C/s.

The microstructure of the plates produced at finish temperatures of accelerated cooling of 610, 500, and 430 °C is shown in Fig. 4.22.

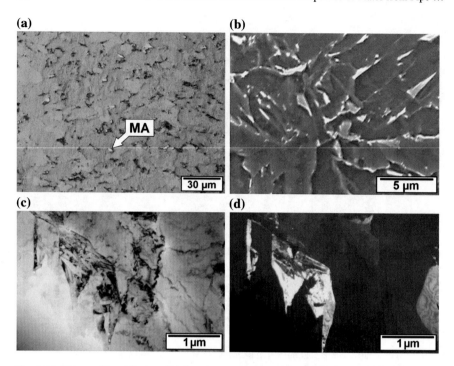

Fig. 4.18 MA constituent (twinned high-carbon martensite with retained austenite): **a** OM (etching in LePera reagent); **b** SEM; **c** TEM (light field); **d** TEM (dark-field image in the austenite reflection)

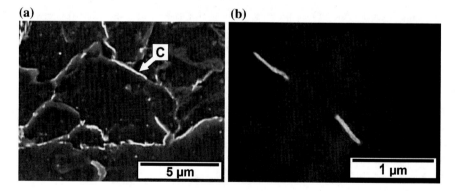

Fig. 4.19 Cementite (C): **a** SEM; **b** TEM (dark-field image in the cementite reflection)

After accelerated cooling finished at a temperature of 610 °C, the microstructure contained quasipolygonal ferrite matrix and regions of degenerate pearlite (Fig. 4.22a).

As the accelerated cooling was finished at lower temperatures of the bainite transformation region ($T_{fc} = 500$ °C), a ferritic–bainitic microstructure consisting of quasypoligonal ferrite and high-carbon bainite was formed (Fig. 4.22b).

Table 4.5 Experimental CR + AC regimes and mechanical properties of the plates 18 mm thick from the 0.05%C-1.25%Mn-0.70%(Cr + Ni + Cu)-(Ti + Nb + V)-steel produced using various T_{fr} and T_{sc}

Treatment	CR + AC regime				Mechanical properties			
	T_{fr} (°C)	T_{sc} (°C)	T_{fc} (°C)	V_c (°C/s)	$\sigma_{0.5}$ (N/mm^2)	σ_B (N/mm^2)	δ_5 (%)	$\sigma_{0.5}/\sigma_B$
Regime I ($T_{fr} > Ar_3$, $T_{sc} > Ar_3$)	$A_{r3} + 60$	$A_{r3} + 20$	540–560	16–18	460	510	26	0.90
Regime II ($T_{fr} > Ar_3$, $T_{sc} < Ar_3$)	$A_{r3} + 15$	$A_{r3} - 20$			460	525	25	0.88
Regime III ($T_{fr} < Ar_3$, $T_{sc} < Ar_3$)	$A_{r3} - 20$	$A_{r3} - 40$			455	515	23	0.88

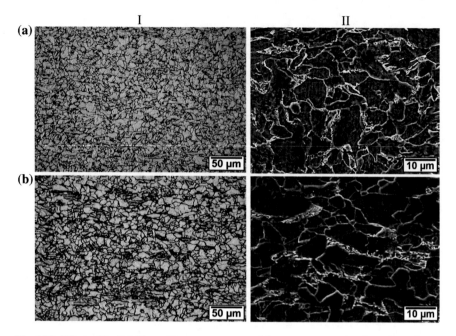

Fig. 4.20 Effect of the finishing rolling temperature and the start temperature of accelerated cooling on the microstructure of the 0.05%C-1.25%Mn-0.70%(Cr + Ni + Cu)-(Ti + Nb + V)-steel plates 18 mm thick, I—OM; II—SEM: **a** regime I ($T_{fr} > Ar_3$, $T_{sc} > Ar_3$); **b** regime III ($T_{fr} < Ar_3$, $T_{sc} < Ar_3$)

Fig. 4.21 Effect of the start temperature of accelerated cooling on the strength properties of the 0.05%C-1.43%Mn-0.16%Mo-0.016%Ti-0.050%Nb-steel plates

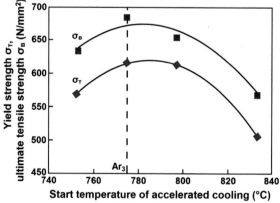

A decrease in T_{fc} to 430 °C leads to the refinement of the high-carbon phase in the form of high-carbon bainite and cementite at grain boundaries (Fig. 4.22c). Figure 4.23 shows the steel structure after etching in a LePera reagent. It is seen that uniformly distributed islands of the MA constituent are also present in the structure.

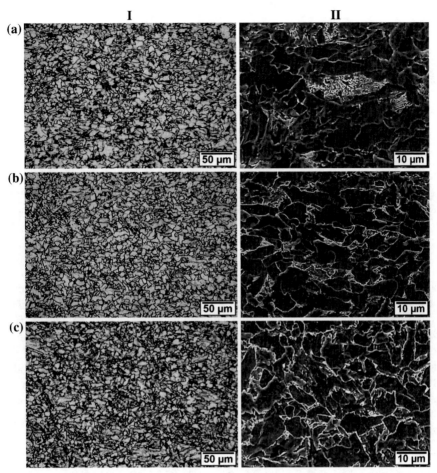

Fig. 4.22 Effect of the finish temperature of accelerated cooling on the microstructure of the 0.07%C-1.33%Mn-0.70%(Cr + Ni + Cu)-(Ti + Nb + V)-steel plates 20 mm thick, I—OM; II—SEM: **a** 610 °C; **b** 500 °C; **c** 430 °C

Fig. 4.23 Microstructure of the 0.07%C-1.33%Mn-0.70%(Cr + Ni + Cu)-(Ti + Nb + V)-steel plate 20 mm thick rapidly cooled to a temperature of 430 °C, OM (etching in a LePera reagent)

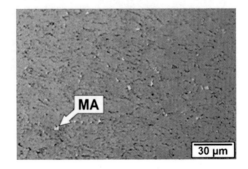

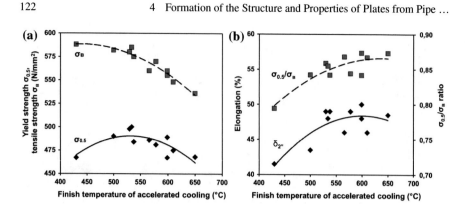

Fig. 4.24 Effect of the finish temperature of accelerated cooling on the strength properties (**a**), relative elongation and $\sigma_{0.5}/\sigma_B$ ratio (**b**) of the 0.07%C-1.33%Mn-0.70%(Cr + Ni + Cu)-(Ti + Nb + V)-steel plates 20 mm thick

The effect of the finish temperature of accelerated cooling on the strength properties, relative elongation, and the $\sigma_{0.5}/\sigma_B$ ratio is shown in Fig. 4.24. As this temperature decreases from 650 to 430 °C, the tensile strength increases by 50 N/mm^2 (Fig. 4.24a). The yield strength exhibits nonmonotone dependence on T_{fc}. As T_{fc} decreases from 650 to 520–540 °C, the yield strength increases by 30 N/mm^2, and a further decrease in T_{fc} to 430 °C leads to a decrease in $\sigma_{0.5}$ by 25 N/mm^2. A decrease in $\sigma_{0.5}$ at low T_{fc} is a consequence of the structural changes associated with the formation of the regions of MA constituent, which leads to the disappearance of the yield plateau in the stress–strain diagram (Fig. 4.25). As T_{fc} decreases within the limits under study, the relative elongation $\delta_{2''}$ decreases from 48 to 42%, and the $\sigma_{0.5}/\sigma_B$ ratio decreases from 0.87 to 0.80 (Fig. 4.24b).

Cooling rate The effect of the cooling rate on the plate microstructure and properties was studied on the 0.06%C-1.35%Mn-0.70%(Cr + Ni + Cu)-(Ti + Nb + V)-steel plates 20 mm thick. The start temperature of the accelerated cooling of the test plates was $Ar_3 + (20–30)$ °C, and the finish temperature of the accelerated cooling was in a range of 520–540 °C. The cooling rate varied between 2 and 25 °C/s.

The ferritic–pearlitic microstructure consisting of polygonal ferrite and lamellar pearlite is formed in the plates cooled in air at a rate of 2 °C/s, (Fig. 4.26a). After cooling at a rate of 15 °C/s, the microstructure is ferritic–bainitic and consists of QPF and UB regions (Fig. 4.26b). As the cooling rate increases to 25 °C/s, the high-carbon structures consisting of UB and cementite particles at grain boundaries of QPF become finer (Fig. 4.26c).

Figure 4.27 shows the effect of the cooling rate on the tensile properties of the plates. As V_c increases from 2 to 15 °C/s, the tensile strength and the yield strength increase by 40 and 60 N/mm^2, respectively (Fig. 4.27a). A further increase in V_c from 15 to 25 °C/s additionally increases σ_B and $\sigma_{0.5}$ by 60 and 35 N/mm^2, respectively. A higher increase in the yield strength compared with the ultimate tensile strength with

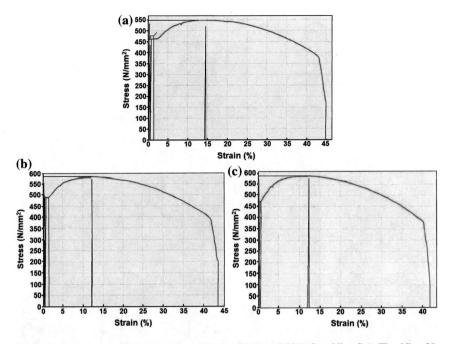

Fig. 4.25 Stress–strain diagrams of the 0.07%C-1.33%Mn-0.70%(Cr + Ni + Cu)-(Ti + Nb + V)-steel plates 20 mm thick cooled to various finish temperature of accelerated cooling: **a** 610 °C; **b** 500 °C; **c** 430 °C

increasing cooling rate from 2 to 15 °C/s is associated with a substantial ferrite grain refinement and the formation of ferritic–bainitic structure instead of the ferritic—pearlitic one. A higher increase in the ultimate tensile strength compared to the yield strength with increasing cooling rate from 15 to 25 °C/s is due to an increase in the fraction of the products of the intermediate transformation without any substantial grain refinement.

Simultaneously with the increase in the strength properties, the relative elongation $\delta_{2''}$ decreases by $\approx 8\%$ with increasing cooling rate, but still remains sufficiently high (Fig. 4.27b). A low $\sigma_{0.5}/\sigma_B$ ratio after cooling at a rate of 2 °C/s is caused by the formation of the structure of non-deformed polygonal ferrite and pearlite. As the cooling rate rises from 15 to 25 °C/s, the $\sigma_{0.5}/\sigma_B$ ratio decreases by 0.03–0.04.

The Charpy V-notch impact energy at a test temperature of 0 °C for all plates under study was in a range of 300–450 J, and the fraction of the ductile constituent in the fracture surface of the specimens was above 90%. No substantial dependences of KV^0 on the CR + AC parameters have been found.

Hard constraints imposed on the carbon content (usually, $\leq 0.08\%$) and on the sulfur content (<0.001%) promote an increased fracture resistance of the steels at subzero temperatures. Figure 4.28 shows the effect of the mass fractions of carbon and sulfur in the 0.06-0.09%C-1.65%Mn-0.80%(Cr + Ni + Cu)-(Ti + Nb)-steel on

I   II

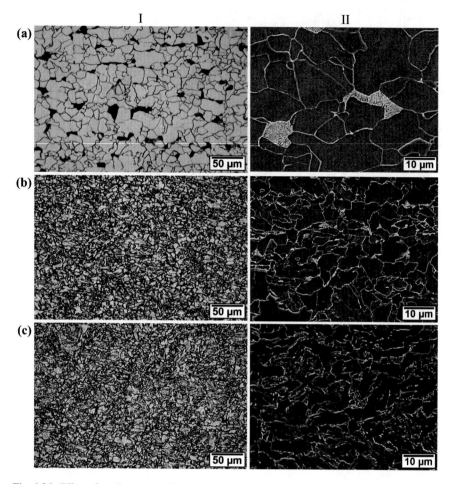

Fig. 4.26 Effect of cooling rate on the microstructure of the 0.06%C-1.35%Mn-0.70%(Cr + Ni + Cu)-(Ti + Nb + V)-steel plates 20 mm thick, I—OM; II—SEM: **a** 2 °C/s; **b** 15 °C/s; **c** 25 °C/s

the Charpy impact energy at a test temperature of −10 °C for X70 grade plates 19 mm thick after processing by the same CR + AC regimes. It is seen that, as the carbon content decreases from 0.09 to 0.06% at S = 0.004%, the average KV^{-10} value increases by 85 J (from 205 to 290 J). As the sulfur content decreases from 0.005 to 0.002% at C = 0.07%, the average KV^{-10} value increases by 75 J (from 230 to 305 J). Simultaneously with the average values, the minimum and maximum values of the impact energy of the specimens increase with decreasing carbon and sulfur contents.

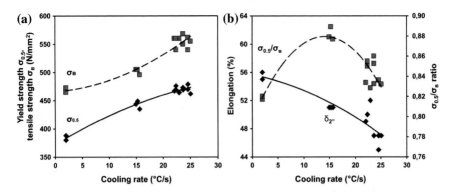

Fig. 4.27 Effect of the cooling rate on the strength properties (**a**), relative elongation, and $\sigma_{0.5}/\sigma_B$ ratio (**b**) of the 0.06%C-1.35%Mn-0.70%(Cr + Ni + Cu)-(Ti + Nb + V)-steel plates 20 mm thick

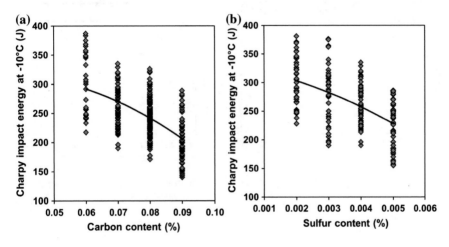

Fig. 4.28 Effects of carbon content at S = 0.004% (**a**) and sulfur content at C = 0.07% (**b**) in the 0.06–0.09%C-1.65%Mn-0.80%(Cr + Ni + Cu)-(Ti + Nb)-steel on the Charpy impact energy at a test temperature of −10 °C for the X70 grade plates 19 mm thick after processing by the same CR + AC regimes

The serial curves of the impact energy and the fraction of the ductile constituent in the fracture surface of Charpy specimens cut from the 0.07%C-1.65%Mn-0.80%(Cr + Ni + Cu)-(Ti + Nb)-steel plates with carbon contents of 0.06 and 0.09% at a sulfur concentration of 0.004% and tested at temperatures ranging from 0 to −70 °C are shown in Fig. 4.29. When the test temperature decreases from 0 to −70 °C, the average Charpy impact energy of the specimens from the plates containing 0.06%C and 0.004%S decreases from 245 to 185 J, and the fraction of the ductile constituent in the fracture surface of the specimens decreases from 100 to 93%. The impact

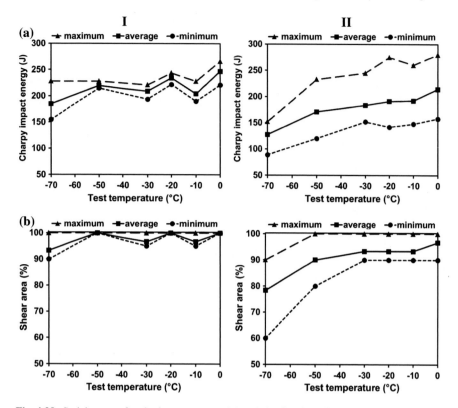

Fig. 4.29 Serial curves for the impact energy (**a**) and the fraction of ductile constituent in the fracture surface (**b**) of Charpy specimens from X70 grade 0.06-0.09%C-1.65%Mn-0.80%(Cr + Ni + Cu)-(Ti + Nb)-steel plate 19 mm thick: I—0.06%C, 0.004%S; II—0.09%C, 0.004%S

energy of the steel plates containing 0.09%C and 0.004%S decreases from 215 to 125 J, and the fraction of the ductile constituent decreases from 97 to 78%. The interval between the minimum and maximum values of the Charpy impact energy and the fraction of the ductile constituent of the steel with 0.09%C was substantially higher than that of the steel with 0.06%C.

No dependence of the shear area in the fracture surface of the DWTT specimens of the plates on the carbon and sulfur contents was revealed for the concentrations under study. The average shear area decreases as the test temperature decreases from 0 to −70 °C, but still remains high (Fig. 4.30).

Fig. 4.30 Serial curves of
the average shear area of
DWTT specimens from the
X70 grade 0.06-0.09%C-
1.65%Mn-0.80%(Cr + Ni +
Cu)-(Ti + Nb)-steel
plate 19 mm thick

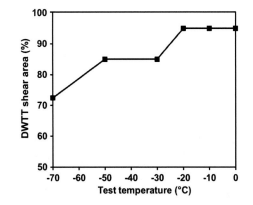

References

Bufalini, P., Pontremoli, M., Aprile, A. (1983). Accelerated cooling after control rolling of line–pipe plates influence of process conditions on microstructure and mechanical properties. In: *Proceedings of the International Conference on Technology and Applications of HSLA steels, Philadelphia, Pennsylvania* (pp. 743–753) October 3–6 1983.

Efron, L. I. (2012). *Metal science in big metallurgy. Pipe steels*. Metallurgizdat, Moscow

Efimov, A. A., Kholodnyy, A. A., Shabalov, I. P., et al. (2017). Structure transformations in the course of cooling of low-carbon low-alloy pipe steels. *Problems of Ferrous Metallurgy and Materials Science, 3,* 49–55.

Kholodnyi, A. A., Matrosov, Y. I., Matrosov, M. Y., & Sosin, S. V. (2016). Effect of carbon and manganese on low-carbon pipe steel hydrogen-induced cracking resistance. *Metallurgist, 60*(1), 54–60.

Matrosov, M. Y., Lyasotskii, I. V., Kichkina, A. A., et al. (2012). Microstructure in low-carbon low-alloy high-strength pipe steel. *Steel in Translation, 42*(1), 84–93.

Matrosov, Y. I., Kholodnyi, A. A., Popov, E. S., et al. (2014a). Influence of thermomechanical processing and heat treatment on microstructure formation and HIC resistance of pipe steel. *Problems of Ferrous Metallurgy and Materials Science, 1,* 98–104.

Matrosov, Y. I., Tskitishvili, E. O., Popov, E. S., Konovalov, G. N., & Kholodnyi, A. A. (2014b). Accelerated cooling after controlled rolling during heavy plate pipe steel manufacture in 3600 mill at the Azovstal metallurgical combine. *Metallurgist, 57*(9), 837–844.

Matrosov, Y. I., Kholodnyi, A. A., Popov, E. S., et al. (2014c). Microstructure and properties of rolled plates from the X52-X65 pipe steels after TMT with accelerated cooling. *Problems of Ferrous Metallurgy and Materials Science, 3,* 53–60.

Ringinen, D. A., Chastukhin, A. V., Khadeev, G. E., & Efron, L. I. (2014). Application of Methods of Processes Simulation and Reproduction in Laboratory Conditions for the Development of Technological Schemes for Thermomechanical Rolling. *Problems of Ferrous Metallurgy and Materials Science, 3,* 28–37.

Salganik, V. M., Chikishev, D. N., Pozhidaeva, E. B., & Nabatchikov, D. G. (2016). Analysis of structural and phase transformations in low-alloy steels based on dilatometric studies. *Metallurgist, 59*(9), 766–773.

Shabalov, I. P., Matrosov, Y. I., Kholodnyi, A. A., et al. (2017). *Steel for gas and oil pipelines resistant to fracture in hydrogen sulphide-containing media*. Moscow: Metallurgizdat.

Chapter 5
Effect of Chemical Composition on the Central Segregation Heterogeneity and HIC Resistance of Rolled Plates

The central segregation chemical and microstructural heterogeneity in low-carbon low-alloy pipe steel plates produced from continuously cast slabs occupies a small region of the plate and does not substantially affect the conventional mechanical properties and cold resistance. However, it causes a discontinuity detected by ultrasonic inspection, decreases mechanical properties in the Z direction, increases the tendency to cracking in the weld zone, and, which is especially important for the pipes for sour service, increases the sensitivity to HIC (Pemov and Nosochenko 2003; Matrosov et al. 2014).

The HIC resistance is affected by not only metallurgical factors such as chemical composition, contents of harmful impurities and nonmetallic inclusions, but also by the character of the microstructure formed at the final stage of plate production. The use of a certain chemical composition and a thermomechanical treatment regime substantially affects not only the final microstructure of the base metal of the plates, but also the degree of segregation chemical and structural heterogeneity of the axial zone (Matrosov et al. 2015; Kholodnyi et al. 2016; Shabalov et al. 2017). For an increase in the HIC resistance of plates, it is necessary to minimize the negative effect of central segregation by controlling the structure of the axial zone. For such control, the chemical composition and the regimes of thermomechanical processing of rolled products should be optimized.

5.1 Central Segregation Heterogeneity in Rolled Plates

The formation of an axial chemical and, as a consequence, microstructural segregation heterogeneity of plate metal is the result of the primary segregation processes occurring upon the crystallization of a continuously cast slab and the secondary segregation caused by the polymorphic transformation occurring in the plates upon

© Springer Nature Switzerland AG 2019
I. Shabalov et al., *Pipeline Steels for Sour Service*, Topics in Mining, Metallurgy and Materials Engineering, https://doi.org/10.1007/978-3-030-00647-1_5

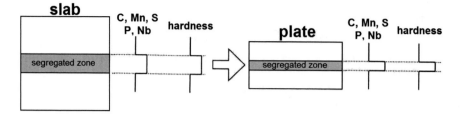

Fig. 5.1 Scheme of the mechanism of central segregation inheritance by the plate made of continuously cast slab

cooling. Figure 5.1 schematically shows the mechanism of the inheritance of the central segregation formed in a slab upon continuous casting by the rolled plate. A local increase in the content of strongly segregating elements (C, Mn, S, P, Nb) and hardness of the axial zone relative to those of the base metal of slab is retained in plate, in which it is concentrated in a narrower zone (Kuznechenko et al. 2017).

Heating of slab for rolling is accompanied by the austenization of its structure and an increase in the diffusion activity of chemical elements in the steel. However, even annealing of slabs for more than 20 h at a temperature of 1250 °C does not substantially affect the distribution of chemical elements over thickness and the HIC resistance of the plates.

The heredity of central segregation from slab to plate can be estimated by the dependence of the microhardness of the axial plate zone on the degree of the initial central segregation of the slab and also by the degree of microstructural heterogeneity over the thickness of the rolled metal. The latter parameter can be expressed by the difference between the microhardnesses of the axial zone and the base metal, $\Delta HV = HV^{az} - HV^{bm}$ or by the coefficient of central segregation structural heterogeneity, $K(HV) = HV^{az}/HV^{bm}$.

Figure 5.2 shows the effect of the rating of central slab segregation according to the Mannesmann scale on the microhardness of the axial zone and the $\Delta HV_{0.2}$ and $K(HV_{0.2})$ parameters of the plates manufactured by two processing schemes such as controlled rolling and controlled rolling with accelerated cooling. It is seen that, regardless of the processing scheme, the microhardness of the axial zone and the degree of structural heterogeneity over the plates thickness increase with increasing central segregation of a continuously cast slab.

The austenite to ferrite transformation occurring upon cooling of the rolled plate is accompanied by secondary segregation processes. Figure 5.3 schematically shows the mechanism of segregation formation in the steel structure upon polymorphic transformation. The hard high-carbon constituents formed in the microstructure of the soft ferrite matrix result from the carbon redistribution occurring upon the $\gamma \rightarrow \alpha$ transformation. At the finish rolling temperature in the austenitic field, carbon is evenly distributed in austenite due to a small size of its atom. Because of its low solubility in ferrite, carbon upon the transformation diffuses into the remaining austenite phase, and this leads to the formation of regions with increased carbon

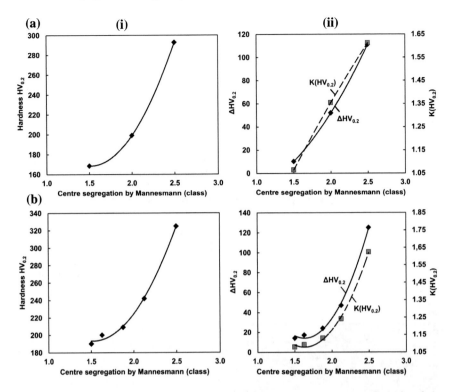

Fig. 5.2 Effect of the rating of central segregation in slabs by the Mannesmann scale on the micro-hardness of the centerline segregation zone (i) and the $\Delta HV_{0.2}$ and $K(HV_{0.2})$ (ii) parameters of the plates manufactured by controlled rolling (**a**) and controlled rolling with accelerated cooling (**b**)

content. Since manganese lowers the $\gamma \rightarrow \alpha$ transformation temperature, the regions with high manganese content undergo transformation in the last turn, and, therefore, their carbon concentration is highest. Thus, the secondary segregation of carbon leads to the formation of a banded ferritic–pearlitic structure, in which the pearlite colonies correspond to the high-manganese regions formed upon crystallization. This manifests itself most clearly in the segregation zone. As a result, the hardenability in such zone increases, and the structure regions of increased hardness are formed, which are favorable for the initiation and propagation of hydrogen-induced cracks.

As is shown by the example of the analysis of chemical composition of the seg-regated martensite band in the steel plate containing 0.11% C, 1.10% Mn, 0.004% P, 0.37% Si, and 0.14% Mo, the segregation of chemical elements in the axial zone of the plate can be measured quantitatively by microprobe (Fig. 5.4) (Usinor Aciers 1987). It is seen that the concentrations of manganese, phosphorus, and molybdenum in the segregation band can be higher by a factor of 1.7, 7, and 2.2, respectively, than in the base metal.

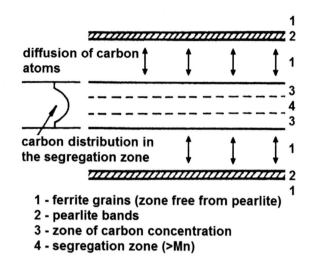

1 - ferrite grains (zone free from pearlite)
2 - pearlite bands
3 - zone of carbon concentration
4 - segregation zone (>Mn)

Fig. 5.3 Scheme of the redistribution of carbon atoms upon the $\gamma \rightarrow \alpha$ transformation with the formation of hard structures

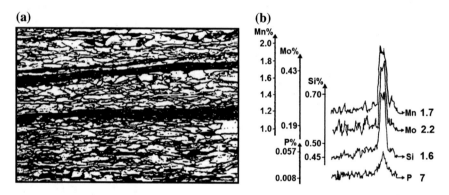

Fig. 5.4 Martensite segregation bands in the centerline segregation zone of rolled plate (**a**) and their chemical composition (**b**) (Usinor Aciers 1987)

The negative effect of the axial segregation on the HIC resistance was revealed on the strips obtained by rolling of the specimens cut from the surface and the center of a continuously cast slab (Fig. 5.5).

The results of the HIC estimation clearly show that the specimens obtained from the surface regions of the slab are more HIC resistant than the specimens made from the central slab zone. This phenomenon is caused by the structure of the segregation bands in the specimens made from the mid-thickness zone of the slab.

Thus, the central segregation chemical heterogeneity formed in continuously cast slabs is inherited by the rolled plates. As a result, the axial zone is the region of the formation of extended hard structures, which are more alloyed than the base metal and have an enhanced tendency to HIC.

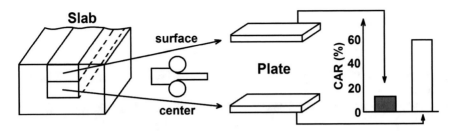

Fig. 5.5 Scheme of the experiment for the determination of the effect central segregation on the HIC resistance of plate (Usinor Aciers 1987)

5.2 Chemical Composition

The chemical composition of steel, especially the content of strongly segregating elements affects the degree of central segregation in the slab and, as a consequence, the segregation heterogeneity in the axial zone of the plates. On the other hand, the chemical composition both in the base metal and in the axial zone affects the kinetics of the polymorphic transformation upon cooling of the plate after rolling, and this also determines the degree of chemical and microstructural heterogeneity of the rolled products. In this section, the effects of the carbon, manganese, and molybdenum contents on the microstructure and properties of the base metal and the central segregation zone and on the HIC resistance of plate are considered.

Carbon Carbon is related to the elements that most strongly segregate in the axial slab zone upon crystallization during casting. Simultaneously with an increase in its content in steel, the susceptibility to primary segregation of elements such as Mn, S, P, and Nb increases (Matrosov et al. 2002). Therefore, the carbon concentration is an important factor affecting the degree of central segregation heterogeneity and the HIC resistance of the plates from low-carbon low-alloy pipe steels (Schwinn and Thieme 2006).

Figure 5.6 shows the macrostructures of the etched polished sections of the plates from steels with different carbon contents. The plates under study were manufactured by regimes of controlled rolling with air cooling.

The central regions of the cross sections of all plates exhibit segregation heterogeneity in the form of highly etched bands parallel to the plate surface. At the same time, the etchability of the axial zone decreases with decreasing carbon content in the steel.

Because of the enrichment of the central segregation zone with the elements affecting the kinetics of the austenite decomposition upon cooling, the structure formed in the axial zone of the plate differs from the structure of the base metal (Fig. 5.7). It is seen that the microstructure of the axial zone, unlike that of the base metal, is characterized by the presence of segregation bands. The degree of segregation heterogeneity decreases with decreasing carbon content. For example, the axial zone of the plate from the steel containing 0.19% carbon exhibits numerous

(a)

(b)

(c)

(d)

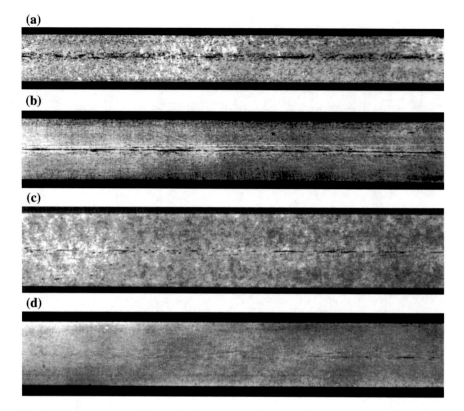

Fig. 5.6 Macrostructure of the polished sections with segregation bands in the central zone of the plates from the steels with different carbon contents: **a**—0.19%C-steel; **b**—0.11%C-steel; **c**—0.08%C-steel; and **d**—0.03%C-steel

pearlite and bainite bands, but the segregation heterogeneity in the 0.03%C-steel is hardly visible.

The distribution of chemical elements C, S, P, Nb, V, and Mn over the thickness of the plates from steels with different carbon contents was studied by an X-ray electron probe microanalysis using a scanning electron microscope equipped with an energy dispersive spectrometer. According to the results of the study, it is established that the chemical segregation in the central zone of the plates decreases with decreasing carbon content in the steel.

Table 5.1 shows the results of the microhardness measurements in the base metal and in the axial zone and the $\Delta HV_{0.2}$ and $K(HV_{0.2})$ parameters for the plates of steels with different carbon contents.

In all cases, the microhardness of the central segregation zone is higher than that of the base metal and depends substantially on the carbon content in the steel. It is seen that the difference in microhardness between the base metal and the central segregation zone decreases from 118 $HV_{0.2}$ to 10 $HV_{0.2}$ as the carbon content decreases

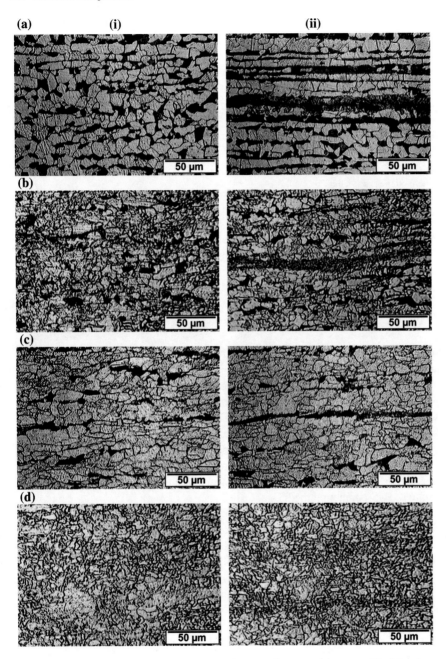

Fig. 5.7 Microstructure of the base metal (i) and the centerline segregation zone (ii) of the plates from the steels with different carbon contents: **a**—0.19%C-steel; **b**—0.11%C-steel; **c**—0.08%C-steel; and **d**—0.03%C-steel

Table 5.1 Microhardness of the base metal and centerline zone and the $\Delta HV_{0.2}$ and $K(HV_{0.2})$ parameters in the plates from the steels with different carbon contents

Steel	Microhardness $HV_{0.2}$		$\Delta HV_{0.2}$	$K(HV_{0.2})$
	Base metal	Axial zone		
0.19%C-steel	149	267	118	1.79
0.11%C-steel	182	293	111	1.61
0.08%C-steel	147	199	52	1.35
0.03%C-steel	158	168	10	1.06

Fig. 5.8 Effect of carbon content on the crack length ratio (CLR)

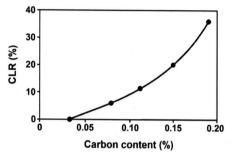

in the range under consideration, from 0.19 to 0.03%. The coefficient of segregation structural heterogeneity $K(HV_{0.2})$ also decreases, from 1.79 to 1.06, which indicates a substantial increase in the structural homogeneity over the plate thickness.

Figure 5.8 shows the effect of carbon content on the hydrogen crack length ratio (CLR), which increases with increasing carbon content in steel. The industrial steels under study were differently alloyed and were rolled by controlled rolling technology with air cooling. The shown dependence of the CLR on the carbon content should be considered as an indication of the need to reduce the carbon content as one of the most important prerequisites for creating steels for large-diameter pipes resistant to failure in H_2S-containing media.

The effect of carbon content changing from 0.04 to 0.08% at comparable manganese contents in a range of 1.25–1.35% on the central segregation microstructural heterogeneity and HIC resistance of the plates 22 mm thick manufactured by controlled rolling followed by accelerated cooling has been studied in (Kholodnyi et al. 2016). The plates were cooled from the austenite field to the lower temperatures of bainitic transformation range at a rate of more than 20 °C/s. The base metal of the plates with a carbon content of 0.04 and 0.08% had a homogeneous ferritic–bainitic microstructure and consisted of a matrix of quasipolygonal ferrite with uniformly distributed regions of high-carbon bainite and cementite particles at the boundaries of ferrite grains (Fig. 5.9).

The carbon content within the limits under study substantially affects the microstructure of the axial zone of the plates (Fig. 5.10). At 0.04% C, a slight structural heterogeneity is seen in the axial zone in the form of a narrow segregation band with fine regions of the MA structure constituent. As the carbon content increases

(a)   **(i)**   **(ii)**

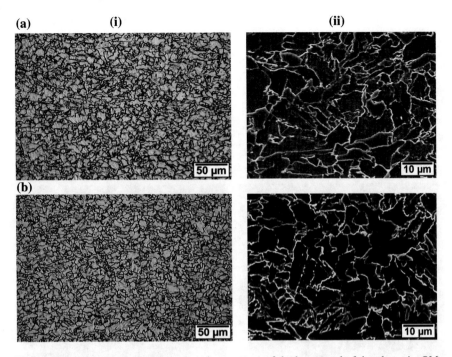

(b)

Fig. 5.9 Effect of carbon content on the microstructure of the base metal of the plates, i—OM; ii—SEM: **a**—0.04% C; **b**—0.08% C

to 0.08%, the central structural heterogeneity becomes more pronounced, which is expressed in an increase in the number of segregation bands containing coarse regions of the MA constituent.

Figure 5.11 shows the microhardness of the base metal and axial zone and the $\Delta HV_{0.2}$ and $K(HV_{0.2})$ parameters of the experimental plates as a function of carbon content.

As the carbon content increases from 0.04 to 0.08%, the microhardness of the base metal slightly increases, from 183 to 200 $HV_{0.2}$. Simultaneously, there is a significant increase in the microhardness of the axial zone, from 200 $HV_{0.2}$ at 0.04% C to 325 $HV_{0.2}$ at 0.08% C. Correspondingly, the difference between microhardnesses of the base metal and the axial zone increases from 17 to 125 $HV_{0.2}$, and the segregation structural heterogeneity coefficient $K(HV_{0.2})$ increases from 1.09 to 1.63.

Figure 5.12 shows the effect of the carbon content in the test steels on the HIC parameters of the plates. The solid lines correspond to the average results of HIC parameter measurements, and the dashed lines correspond to their limiting values.

A decrease in the carbon content, other things being equal, resulted in a substantial increase in the HIC resistance of the plates, i.e., in a decrease in the HIC parameters. The CLR parameter on average decreased from 12.5% at 0.08% C to zero at 0.04% C.

Hydrogen cracks propagated in the plate axial zone along the segregation bands of increased hardness with the MA constituents (Fig. 5.13).

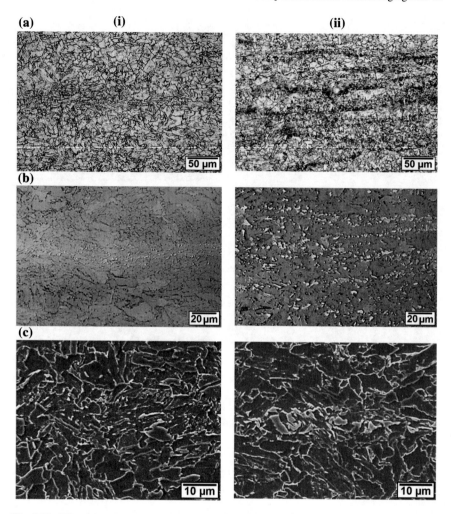

Fig. 5.10 Effect of carbon content on the microstructure of the centerline segregation zone of the plates, **a**—OM; **b**—OM (etching in LePera reagent); **c**—SEM: i—0.04% C; ii—0.08% C

Figure 5.14 shows the microstructure of the axial zone and the chemical composition of the segregation band of the plates from the steels containing 0.01 and 0.07% carbon (0.25% Si, 1.50% Mn, and 0.020% P) (Ohtani et al. 1983). It is seen that an increase in the carbon content promotes the formation of pronounced segregation chemical and structural heterogeneity in the axial zone of the plate.

Even at a relatively low carbon content (0.025–0.060%), a decrease in CSR has been revealed in (Barthold et al. 1989) with decreasing mass fraction of carbon in the X65 grade steel (Fig. 5.15).

The experience shows that, to increase the HIC resistance of steel, one should maintain the carbon content at a level of maximum 0.07%.

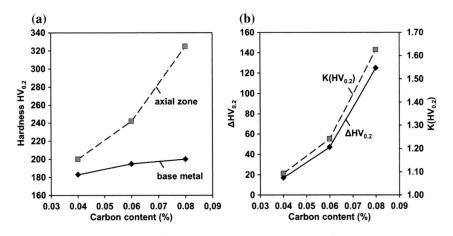

Fig. 5.11 Effect of carbon content on the microhardness of the base metal and the centerline segregation zone (**a**) and the $\Delta HV_{0.2}$ and $\underline{K}(HV_{0.2})$ parameters (**b**) of the plates

Manganese Manganese is the main alloying element in low-alloy pipe steels. Depending on the pipe application and grade, its content is usually in a range of 1.40–1.90%. However, for steels resistant to cracking in H_2S-containing media, its content is maintained at a lower level (usually no higher than 1.30%), which is caused by its increased tendency to segregation upon solidification of liquid metal during continuous casting.

The effect of manganese concentration on the central structural heterogeneity and the HIC susceptibility of plates was studied in (Kholodnyi et al. 2016). The steel contained from 0.65 to 1.35% manganese at 0.06% carbon. Plates 15–25 mm thick were divided into two groups in accordance with the technological regimes of accelerated cooling after controlled rolling:

- regime No. 1: $T_{sc} \geq Ar_3$, $T_{fc} = 540$–$640\,°C$, $V_c = 12$–$18\,°C/s$;
- regime No. 2: $T_{sc} \geq Ar_3$, $T_{fc} = 490$–$530\,°C$, $V_c = 22$–$26\,°C/s$.

The plates were rapidly cooled from the single-phase γ field at both treatment regimes, but regime No. 1 used a higher finish temperature of accelerated cooling and a lower cooling rate. The estimation of the effect of manganese on the susceptibility to HIC of the plates processed by the above regimes (Fig. 5.16) shows that, in both cases, the reduction in manganese content positively affects the HIC resistance of the plates from the test steels. At the same time, a decrease in the finish temperature of the accelerated cooling and an increase in cooling rate (plates processed by regime No. 2) substantially increase the HIC resistance compared to that of the plates processed by regime No. 1. For example, at a manganese content of 1.00%, the treatment regime No. 1 provided an average CLR of 5% (maximum 12%), while after the treatment by regime No. 2, CLR was 0.5% (maximum 1.6%). On average, the HIC parameters of the plates treated by regime No. 2 were smaller by a factor of about eight than those of the plates treated by regime No. 1 in the entire range of manganese contents used.

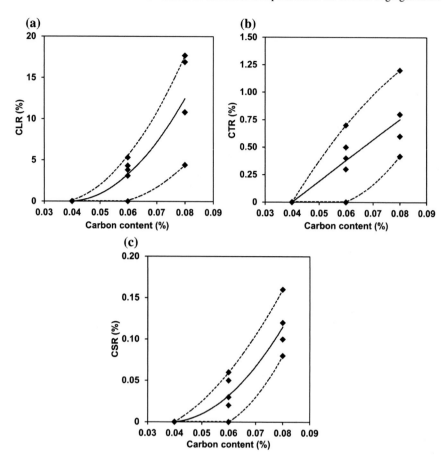

Fig. 5.12 Effect of carbon content on the HIC parameters of the plates: **a**—CLR; **b**—CTR; **c**—CSR

After treatment by regime No. 2, the HIC parameter reached zero at a manganese content of about 0.90%. Thus, the additive effect of a decrease in manganese content and a simultaneous decrease in the finish temperature of the accelerated cooling and an increase in the cooling rate has been established for an increase in the HIC resistance of the plates.

The microstructure of the experimental plates containing 0.65 and 1.35% manganese after treatment by regime No. 2 contained the quasipolygonal ferrite matrix with uniformly distributed regions of high-carbon bainite and cementite (Fig. 5.17).

A slight structural heterogeneity in the form of segregation bands with fine uniformly distributed regions of high-carbon bainite and MA constituent was observed in the axial zone of the plate containing 0.65% manganese (Fig. 5.18). As the manganese content increases to 1.35%, the degree of central structural heterogeneity increases. Coarse regions of the MA constituent are present in the segregation bands.

(a)

(b)

(c)

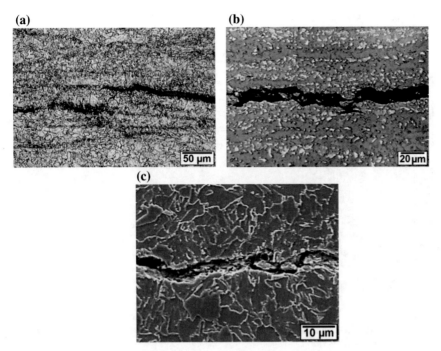

Fig. 5.13 Hydrogen-induced cracks in the centerline segregation zone of the plate containing 0.08% carbon: **a**—OM; **b**—OM (etching in LePera reagent); **c**—SEM

With increasing manganese content from 0.65 to 1.35%, the microhardness of the axial zone increased by 52 $HV_{0.2}$ (from 190 to 242 $HV_{0.2}$), while the microhardness of the base metal increased by only 19 $HV_{0.2}$ (from 176 to 195 $HV_{0.2}$) (Fig. 5.19). A consequence of such change in the microhardness of the zones over the plate thickness, the $\Delta HV_{0.2}$ parameter increased from 14 to 47 $HV_{0.2}$ and the $K(HV_{0.2})$ ratio increased from 1.08 to 1.24.

The hydrogen-induced cracks propagated in the axial zone of the plates along the interfaces between ferrite and MA constituent in the segregation bands (Fig. 5.20).

As a result of the study, it was established that a decrease in the mass fraction of manganese increases the homogeneity of the plate microstructure over thickness and, thus, increases their HIC resistance.

The degree of manganese segregation strongly depends on the total manganese content in steel (Fig. 5.21a). The segregation of manganese abruptly increases at its content of more than 1.00%. This is also confirmed by the microhardness measurement in the segregation band in plates with different manganese contents (Fig. 5.21b).

Figure 5.22 shows the maximum hardness in the middle and at ¼ of the thickness of the plates from the low-sulfur calcium-treated steel with different manganese contents.

0.01%C-1.5%Mn-0.02%P

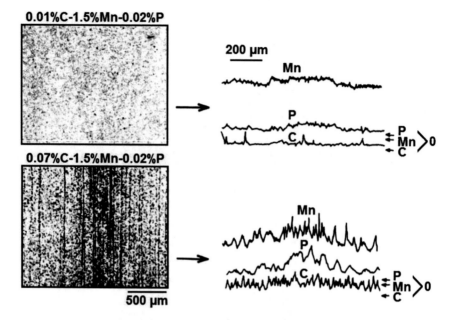

Fig. 5.14 Effect of the carbon content on the central segregation chemical and structural hetero-geneity (Ohtani et al. 1983)

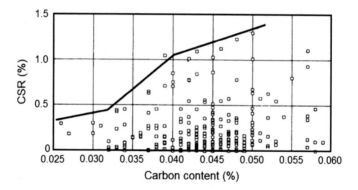

Fig. 5.15 Effect of carbon content on the CSR parameter upon test of plates in Solution A (Barthold et al. 1989)

At a manganese content of 1.20%, the hardness in the middle of the plate was slightly higher than at 1/4 of the plate thickness. On the contrary, at 1.50% Mn, the microhardness of the axial zone was sufficiently high, of about 300 HV, while at ¼ plate thickness it was below 240 HV. Figure 5.22 also shows the boundary microhardness (HV 260), above which the steel becomes sensitive to HIC. Therefore, the central zone of the plate from the steel containing 1.5% Mn is very sensitive to HIC.

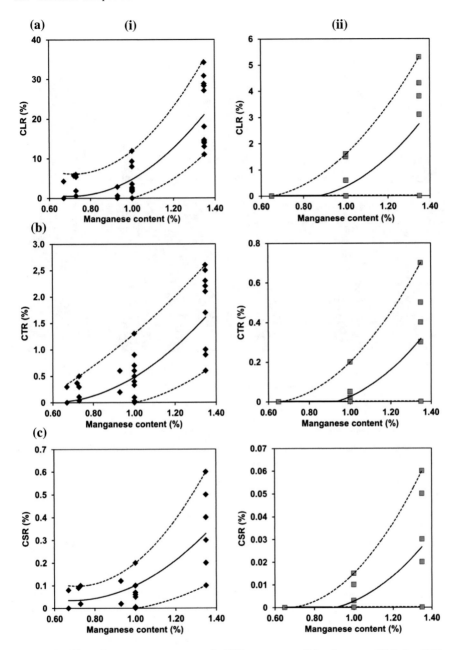

Fig. 5.16 Effect of manganese content on the HIC parameters of the plates, **a**—CLR; **b**—CTR; **c**—CSR: i—regime No. 1—$T_{sc} \geq Ar_3$, $T_{fc} = 540$–$640\ °C$, $V_c = 12$–$18\ °C/s$; ii—regime No. 2—$T_{sc} \geq Ar_3$, $T_{fc} = 490$–$530\ °C$, $V_c = 22$–$26\ °C/s$

(a)   **(i)**   **(ii)**

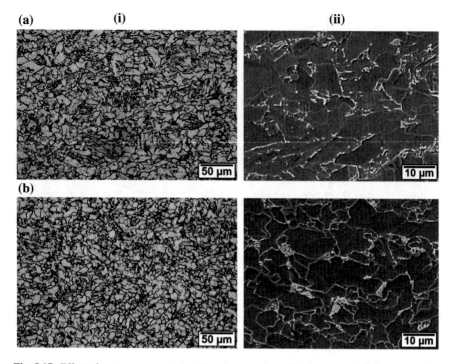

Fig. 5.17 Effect of manganese content on the microstructure of the base metal of plates treated by regime No. 2, i—OM; ii—SEM: **a**—0.65% Mn; **b**—1.35% Mn

Figure 5.23 presents the micrograph of a hydrogen-induced crack in the central segregation zone of the steel containing 0.05%C-1.5%Mn-0.0005%S. The steel microstructure was mainly bainitic. However, the central zone contains lower bainite and MA inclusions. This strengthened zone was formed due to the segregation of alloying elements such as manganese and carbon. Cracking in the central zone of the plate can be prevented by reducing Mn content in the low-carbon low-sulfur Ca-treated steel.

Figure 5.24 shows the effect of manganese content on the CLR and CSR parameters of the plates with different carbon contents after treatment by different regimes. A substantial decrease in HIC resistance of the plates containing 0.05–0.15% C in the state after controlled rolling is observed as the manganese content increases above 1.20%. At Mn $\leq$ 0.90%, the CLR and CSR parameters remain virtually unchanged. At an ultra-low carbon content of 0.02%, the CLR and CSR parameters are substantially smaller than at 0.05–0.15% C even if the manganese content ranges between 1.65 and 2.00%. Quenching and tempering of the plates with an increased carbon content even at a manganese content of more than 1.60% allowed a substantial reduction in the HIC parameters compared with those realized after the processing by controlled rolling technology.

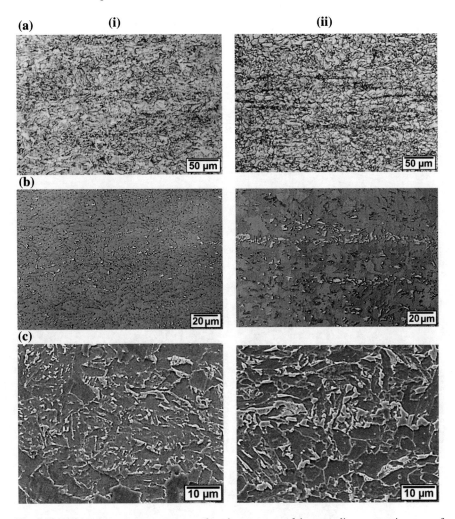

Fig. 5.18 Effect of manganese content on the microstructure of the centerline segregation zone of plates treated by regime No. 2: **a**—OM; **b**—OM (etching in LePera reagent); **c**—SEM: i—0.65% Mn; ii—1.35% Mn

Many authors note the need to reduce the manganese content below 1.20% for the reduction in the central segregation and increase in the HIC resistance of steel plates, especially from the slabs continuously cast with out-date CCM. For the reason of a strong tendency of manganese to segregation, a concept of low-manganese (0.20–0.30%) steel is proposed with good strength characteristics provided by high chromium content (0.40–0.55%), alloying with copper and nickel, and high niobium content (0.075–0.095%).

In low-carbon low-alloy pipe steels, the carbon and manganese contents substantially affect the mechanical properties of rolled plates and cannot be reduced to

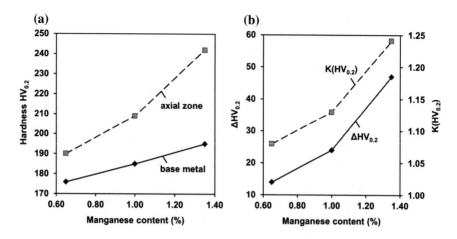

Fig. 5.19 Effect of manganese content on the microhardness of the base metal and the centerline segregation zone (**a**) and the $\Delta HV_{0.2}$ and $K(HV_{0.2})$ (**b**) parameters of the plates treated by regime No. 2

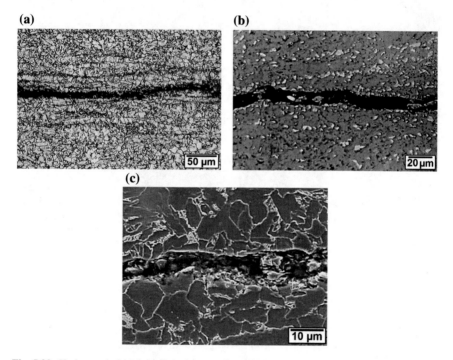

Fig. 5.20 Hydrogen-induced cracks in the centerline segregation zone of the plate with a manganese content of 1.35% after treatment by regime No. 2: **a**—OM; **b**—OM (etching in LePera reagent); **c**—SEM

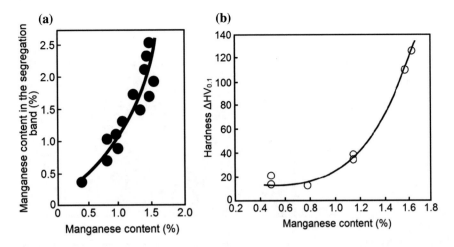

Fig. 5.21 Effect of the total content of manganese on its concentration in the segregation band (**a**) and the difference between the microhardnesses of the base metal and the segregation zone (**b**) (Usinor Aciers 1987)

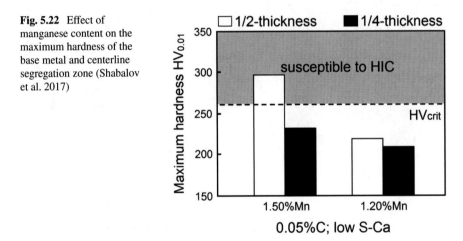

Fig. 5.22 Effect of manganese content on the maximum hardness of the base metal and centerline segregation zone (Shabalov et al. 2017)

extremely low values. As their contents are limited, additional expensive alloying (Cr, Ni, Cu, and Mo) and microalloying (Nb and V) elements are needed to provide the specified strength properties of the plates. When developing the chemical composition of the steel for sour service, one should maintain the mass fraction of C, Mn, and alloying elements at a balanced level, at which of the required mechanical properties and increased HIC resistance can be simultaneously provided without significant increase in production costs. Therefore, the effect of the carbon and manganese contents on the central segregation and HIC resistance of plates is often considered together.

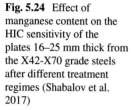

Fig. 5.23 Hydrogen-induced crack in the central segregation zone of the plate from the steel containing 0.05% C-1.50% Mn-0.0005% S (Shabalov et al. 2017)

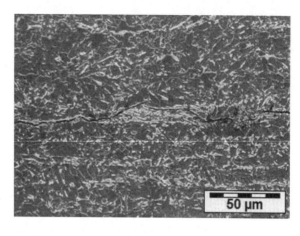

Fig. 5.24 Effect of manganese content on the HIC sensitivity of the plates 16–25 mm thick from the X42-X70 grade steels after different treatment regimes (Shabalov et al. 2017)

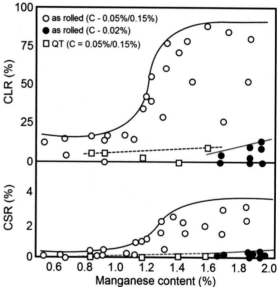

Figure 5.25 shows the effects of carbon content changing in a range of 0.01–0.15% and manganese content changing in a range of 1.50–2.50% on the microstructure of the axial zone of the plates.

It is seen that some changes in the contents of both carbon and manganese substantially affect the degree of the microstructural heterogeneity in the axial zone of the plates. At the minimum carbon and manganese concentrations under consideration, 0.01 and 1.50%, respectively, there is virtually no segregation heterogeneity. In this case, a separate or simultaneous increase in the carbon and/or manganese contents leads to an increase in the degree of central segregation heterogeneity.

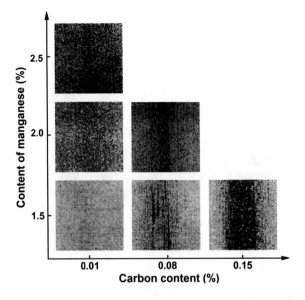

Fig. 5.25 Effect of the carbon and manganese contents on the central segregation structural heterogeneity of the plates (Ohtani et al. 1983)

The acceptable manganese content providing the HIC resistance of plates depends on the carbon content in the steel (Efron 2012). The higher is the carbon content, the lower is the permissible manganese content. Figure 5.26 shows the regions of favorable ratios between carbon and manganese. The maximum allowable carbon content depends on the manganese concentration in the steel and on the production technology: The limiting manganese content for HIC at a specific carbon content in the steel is higher after quenching and tempering than in the hot rolled state.

The data given in Fig. 5.27 show that a simultaneous decrease in manganese content in segregation zone and carbon contents suppresses the HIC development. It is seen that, at a hardness of the segregation zone below 300 HV, quite satisfactory HIC resistance is achieved (Usinor Aciers 1987).

At comparable manganese contents, the yield ratio by a criterion of CLR ≤ 15% was 90% at a carbon content of 0.03–0.06% and about 12% at a higher carbon content of 0.09–0.12% (Fig. 5.28) (Barthold et al. 1989).

Thus, to reduce the central segregation heterogeneity in the axial zone of the plates and to increase the HIC resistance of the steel, both the carbon and manganese contents should be corrected simultaneously.

Molybdenum The role of molybdenum in the formation of the structure and properties of high-strength low-alloy pipe steels manufactured by the technology of thermomechanical processing is well known. Molybdenum, which is a relatively expensive element, is nevertheless used in pipe steels for the achievement of high strength properties (for steels of X80 (K65) and higher grades) (Efron 2012; Morozov and

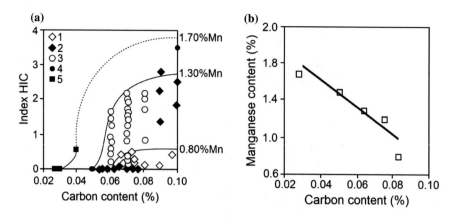

Fig. 5.26 Effect of carbon and manganese contents on HIC resistance of plates; **a**—manganese content: 1—0.80%; 2—1.20%; 3—1.30%; 4—1.50%; 5—1.70%, **b**—region of favorable ratios between carbon and manganese (below the line) (Efron 2012)

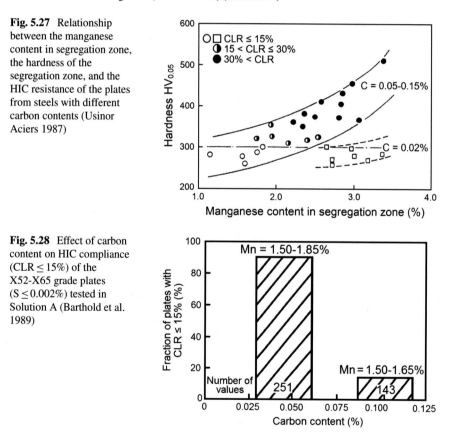

Fig. 5.27 Relationship between the manganese content in segregation zone, the hardness of the segregation zone, and the HIC resistance of the plates from steels with different carbon contents (Usinor Aciers 1987)

Fig. 5.28 Effect of carbon content on HIC compliance (CLR ≤ 15%) of the X52-X65 grade plates (S ≤ 0.002%) tested in Solution A (Barthold et al. 1989)

Naumenko 2009) or for the production of thick plates ($\geq$25 mm) for underwater gas pipelines (Gray 2011, Il'inskii et al. 2014). It was of interest to study the possibility of simultaneous enhancement of strength properties and realization of high cracking resistance of the plates from pipe steels in hydrogen sulfide-containing media by using small molybdenum additions. The effect of 0.15% molybdenum on the microstructure and properties of microalloyed pipe steel was studied in (Kholodnyi et al. 2017) on the test steels with the same contents of basic chemical elements. However, the Mo-free-steel was free from molybdenum, while the 0.15%Mo-steel steel contained 0.15% molybdenum (Table 5.2).

Both steels were characterized by high purity of harmful impurities (S = 0.001% and P = 0.010%), and the nonmetallic inclusions. Plates 20 mm thick were thermomechanically processed with a double-stand reversing plate mill 3600 by the controlled rolling technology followed by cooling from the austenite field at a rate of more than 20 °C/s to a temperature ranging from 410 to 565 °C.

Microstructure of the test plates was a matrix of quasipolygonal ferrite (QPF) with uniformly distributed islands of high-carbon bainite (UB) and cementite (C) particles at the ferrite boundaries (Figs. 5.29 and 5.30).

Figure 5.31 shows the effect of the finish temperature of accelerated cooling on the strength properties of the plates. As T_{fc} decreases from 565 to 420 °C, the average yield strength increases by 30 N/mm^2, from 440 to 470 N/mm^2 for the steel without Mo and from 460 to 490 N/mm^2 for the steel with 0.15%Mo.

Table 5.2 Chemical composition of the steels

Steel	Elements (wt%)									
	C	Mn	Si	S	P	Cr	Ni	Cu	Mo	Ti+Nb+V
Mo-free-steel	0.06	0.90–0.95	0.20–0.23	0.001	0.010	0.25	0.25	0.20	–	$\leq$0.12%
0.15%Mo-steel									0.15	

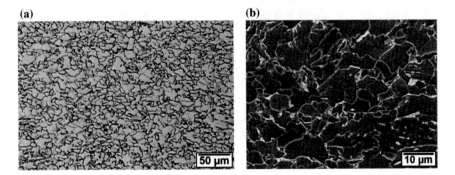

Fig. 5.29 Microstructure of the 0.15%Mo-steel plate after accelerated cooling to $T_{fc} = 420$ °C: **a**—OM; **b**—SEM

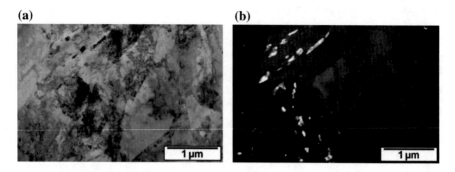

Fig. 5.30 Region of high-carbon bainite and quasipolygonal ferrite in the microstructure of the 0.15%Mo-steel plate after accelerated cooling to $T_{fc} = 420$ °C: **a**—TEM (bright-field image); **b**—TEM (dark-field image in the cementite reflection)

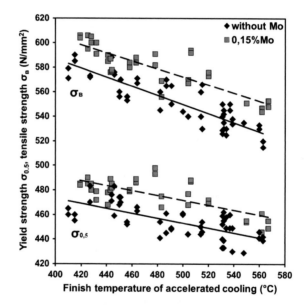

Fig. 5.31 The effect of the finish temperature of accelerated cooling on the yield strength and ultimate tensile strength of plates from the steel Mo-free-steel and 0.15%Mo-steel

In this case, the increase in the ultimate tensile strength was 50–55 N/mm², from 525 to 580 N/mm² and from 550 to 600 N/mm² for steels without Mo and with 0.15% Mo, respectively.

The average yield strength and ultimate tensile strength of the plates from the steel with 0.15% Mo were by about 20 N/mm² higher than those of the steel without Mo.

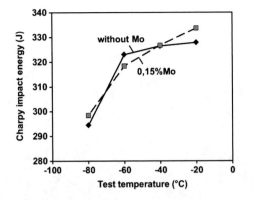

Fig. 5.32 Average Charpy impact energy of the plates from the Mo-free-steel and 0.15%Mo-steel at various test temperatures

Simultaneously with the decrease in the finish temperature of the accelerated cooling, the relative elongation $\delta_{2''}$ insignificantly decreases, by about 3% (from 50 to 47%) for the steel without Mo and by about 4% (from 53 to 49%) for the steel with 0.15% Mo. The average relative elongation of the plates of steel with 0.15% Mo was lower by ~2–3% than those of the steel without Mo. The $\sigma_{0.5}/\sigma_B$ ratio for both steels was at the same level and decreased from ~0.83 to 0.81 as T_{fc} decreased within the range under consideration.

Figure 5.32 shows the results of serial impact bending tests of the plates from the test steels. The Charpy impact energy of the specimens tested at temperatures ranging from -20 to $-80\,^\circ$C was at a comparable level for both steels. An average decrease in the impact energy with decreasing test temperature was 35 J (from ~330 to ~295 J). The impact energy was at least 250 J, even at a test temperature of $-80\,^\circ$C. The test plates exhibited a high resistance to brittle fracture upon DWTT at temperatures decreasing from 0 to $-20\,^\circ$C, the shear area in the fracture surface of the DWTT specimens was 90–100%.

Figure 5.33 shows the microstructure of the centerline segregation zone of plates from the test steels after accelerated cooling to temperatures of 550 and 420 $^\circ$C. In the axial zone of the Mo-free steel plate rapidly cooled to 550 $^\circ$C (Fig. 5.34) and in the axial zone of the plates from the steel with 0.15% Mo after cooling to 550 and 420 $^\circ$C, the segregation bands contained regions of twinned high-carbon martensite with retained austenite (MA constituent).

The segregation bands consisting of coarse packets of acicular bainitic ferrite with interlayers of retained austenite are present at the lath boundaries in the axial zone of the Mo-free-steel plate rapidly cooled to 420 $^\circ$C, (Fig. 5.35).

Such change in the microstructure of the segregation bands in the plates produced at different finish temperatures of accelerated cooling affected their HIC

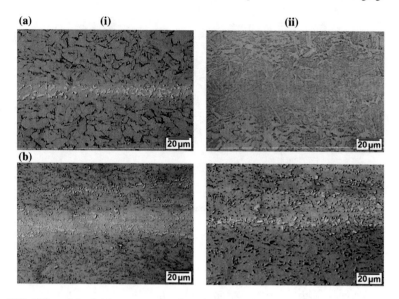

Fig. 5.33 Effect of the finish temperature of accelerated cooling on the microstructure of the centerline segregation zone of the Mo-free-steel (**a**) and 0.15%Mo-steel (**b**) steel plates, OM (etching in LePera reagent): i—$T_{fc} = 550\ °C$; ii—$T_{fc} = 420\ °C$

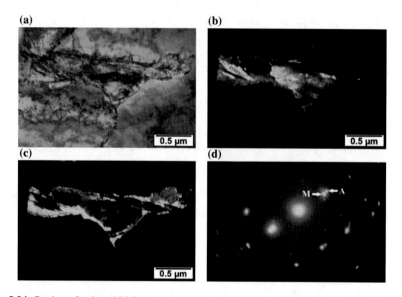

Fig. 5.34 Region of twinned high-carbon martensite with retained austenite (MA constituent) in the segregation band of the Mo-free-steel plate after accelerated cooling to $T_{fc} = 550\ °C$, TEM: **a**—bright-field image; **b**—dark-field image in the martensite reflection; **c**—dark-field image in the austenite reflection; **d**—electron-diffraction pattern of the region shown in (**a**–**c**)

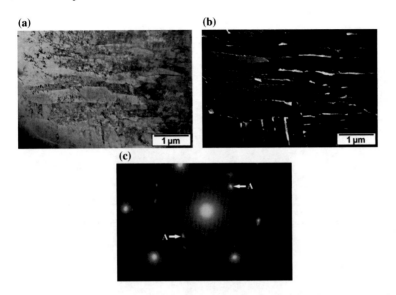

Fig. 5.35 Acicular bainitic ferrite with interlayers of retained austenite at the lath boundaries in the segregated band of the Mo-free-steel plate after accelerated cooling to $T_{fc} = 420$ °C, TEM: **a**—bright-field image; **b**—dark-field image in the austenite reflection; **c**—electron-diffraction pattern from the region shown in (**a, b**)

resistance (Fig. 5.36). A decrease in the finish temperature of accelerated cooling of the molybdenum-free steel plates from 565 to 420 °C deteriorates their HIC resistance as follows: CLR increases from 0% to about 17%, CTR increases from 0% to ~0.58%, and CSR increases from 0% to ~0.12%. The molybdenum-containing plates rapidly cooled to T_{fc} ranging from 560 to 420 °C did not exhibit any substantial change in the HIC resistance. The average CLR parameter of the steel plates with 0.15% Mo after accelerated cooling to different T_{fc} in the range under consideration was about 1%.

The microstructure of the axial zone of the plates after the HIC test is shown in Fig. 5.37. It is seen that the hydrogen-induced cracks propagate in the axial zone along the segregation bands. The plates with segregation bands in the axial zone with regions of the MA constituent exhibited a substantially higher HIC resistance than the plates with segregation bands consisting of coarse packets of acicular bainitic ferrite with interlayers of retained austenite at the lath boundaries.

Thus, the addition of 0.15% Mo into the steel produced by the CR+AC technology increases the HIC resistance of the plates as the finish temperature of the accelerated cooling decreases from 560 to 420 °C.

Simultaneously, there is a substantial increase in strength properties: the yield strength increases by 50 N/mm^2, and the ultimate tensile strength increases by 75 N/mm^2.

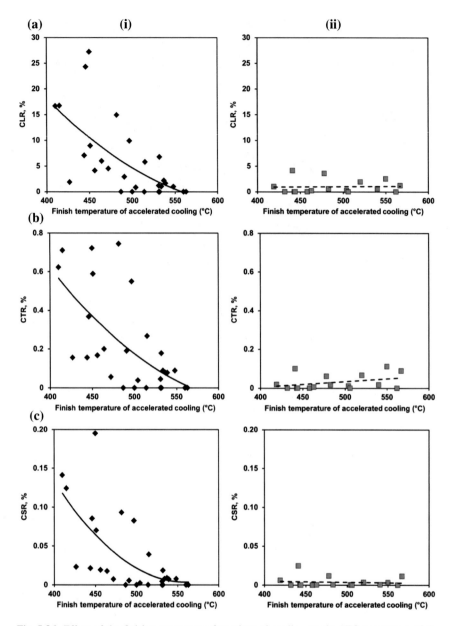

Fig. 5.36 Effect of the finish temperature of accelerated cooling on the HIC parameters of the plates from the Mo-free-steel (i) and 0.15%Mo-steel (ii): **a**—CLR; **b**—CTR; **c**—CSR

(a)   **(i)**   **(ii)**

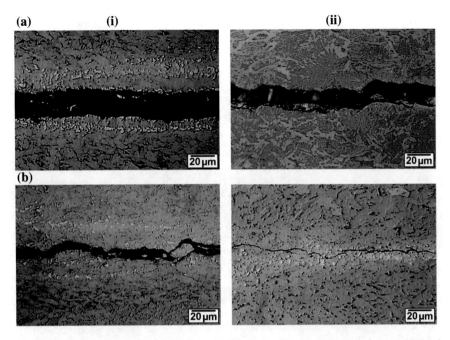

Fig. 5.37 Hydrogen-induced cracks in the centerline segregation zone of the Mo-free-steel (**a**) and 0.15%Mo-steel (**b**) plates, OM (etching in LePera reagent): I—$T_{\text{fc}} = 550\ ^{\circ}$C; II—$T_{\text{fc}} = 420\ ^{\circ}$C

References

Barthold, G., Streißelberger, A., & Bauer, J. (1989). Modern line pipe steels for sour service—Experience in applying TM-rolled and accelerated cooling plate. *Technical information Dillinger Huttenwerke, 9,* 1–8.

Efron, L. I. (2012). *Metal science in big metallurgy pipe steels.* Moscow: Metallurgizdat.

Gray, J. M. (2011). Application of niobium–molybdenum strengthening mechanisms in high strength linepipe steels. In: *Proceedings of the First International Symposium on Fundamentals and Application of Mo and Nb Alloying in High Performance Steels,* Taipei, Taiwan, 7–8 November 2011.

Il'inskii, V. I., Matrosov, M. Y., Stepanov, P. P., et al. (2014). Experience of mastering plate production of strength category SAWL 450 for deep-water pipes at the Vyksa metallurgical plant 5000 mill. *Metallurgist, 58*(1), 38–42.

Kholodnyi, A. A., Matrosov, Y. I., Matrosov, M. Y., & Sosin, S. V. (2016). Effect of carbon and manganese on low-carbon pipe steel hydrogen-induced cracking resistance. *Metallurgist, 60*(1), 54–60.

Kholodnyi, A. A., Matrosov, Y. I., & Sosin, S. V. (2017). Influence of molybdenum on microstructure, mechanical properties and resistance to hydrogen induced cracking of plates from pipe steels. *Metallurgist, 61*(3), 230–237.

Kuznechenko, Y. S., Shabalov, I. P., Kholodnyy, A. A., et al. (2017). Centerline segregation inhomogeneity and resistance to hydrogen induced cracking of rolled plates from pipe steels. Part 1. Influence of chemical composition. *Problems of Ferrous Metallurgy and Materials Science, 2,* 45–57.

Matrosov, Y. I., Kholodnyi, A. A., Matrosov, M. Y., et al. (2015). Effect of accelerated cooling parameters on microstructure and hydrogen cracking resistance of low-alloy pipe steels. *Metallurgist, 59*(1), 60–68.

Matrosov, Y. I., Kholodnyi, A. A., Popov, E. S., et al. (2014). Influence of thermomechanical processing and heat treatment on microstructure formation and HIC resistance of pipe steel. *Problems of Ferrous Metallurgy and Materials Science, 1,* 98–104.

Matrosov, Y. I., Kolyasnikova, N. V., Nosochenko, A. O., & Ganoshenko, I. V. (2002). Influence of carbon and central segregational inhomogeneity on the H_2S resistance of continuous-cast tube steel. *Steel in Translation, 32*(11), 69–74.

Morozov, Y. D., & Naumenko, A. A. (2009). Study of the effect of chemical compound composition on a set of mechanical properties and microstructure of sheet rolled product of strength class K65 (X80). *Metallurgist, 53*(11), 685–692.

Ohtani, H., et al. (1983). Development of low P_{cm} high grade line pipe for Artic service and sour environment. In: *Proceedings of the International Conference on Technology and Applications of HSLA Steels*, Philadelphia, Pennsylvania, pp. 843–854, 3–6 October 1983.

Pemov, I. F., & Nosochenko, O. V. (2003). Improving the mechanical properties of rolled plates and slabs in the thickness direction. *Metallurgist, 47*(11), 460–464.

Schwinn, V., & Thieme, A. (2006). TMCP steel plates for sour service linepipe application. In: *International Seminar "Pipe Seminar Modern Steels for Gas and Oil Transmission Pipelines, Problems and Prospects"*, Moscow (p. 272). Moscow: Metallurgizdat, 15–16 March 2006.

Shabalov, I. P., Matrosov, Y. I., Kholodnyi, A. A., et al. (2017). *Steel for gas and oil pipelines resistant to fracture in hydrogen sulphide-containing media*. Moscow: Metallurgizdat.

Usinor Aciers. (1987). *Steel grades for the manufacture of welded pipes, resistant to cracking under the influence of hydrogen sulphide*. Document of the Working Group on the Metallurgical Industry of Franco-Soviet Cooperation, p. 51.

Chapter 6
Effect of Thermomechanical Treatment on the Central Segregation Heterogeneity and HIC Resistance of Rolled Plates

For the development of the thermomechanical and heat treatment regimes for HIC-resistant plates from pipe steels, their effect on the formation of the base metal microstructure responsible for the mechanical properties of the steel and on the axial zone microstructure responsible for the HIC resistance of the plates is taken into account. The structural formation in the central segregation zone and in the base metal of the plates is mainly affected by the conditions of hot plastic deformation and cooling. The microstructural formation in the centerline zone is characterized by some specific features associated, first of all, with a higher content of segregating elements, which is inherited from continuously cast slabs. This chapter is devoted to the effect of the controlled rolling and post-deformation cooling (cooling rate, start and finish temperature of accelerated cooling) and heat treatment on the microstructure of the central segregation zone and the HIC resistance of rolled plates.

6.1 Deformation-Thermal Treatment

The results of the studies aimed at the development of the thermomechanical treatment scheme for the realization of the axial zone microstructure providing an increased HIC resistance are considered in (Shabalov et al. 2017; Matrosov et al. 2014; Kuznechenko et al. 2017). Plates 14–15 mm thick from the steel containing 0.06%C, 0.23%Si, 0.95%Mn, 0.35%(Ni + Cu), Ti + Nb + V, 0.001%S and 0.010%P were used for the study. The plates were manufactured using various thermomechanical processing options:

- high-temperature controlled rolling with the end of deformation in the lower region of the γ field and cooling in air (HCR);
- low-temperature controlled rolling with the end of deformation in the $(\gamma + \alpha)$ field and cooling in air (LCR);

© Springer Nature Switzerland AG 2019
I. Shabalov et al., *Pipeline Steels for Sour Service*, Topics in Mining, Metallurgy and Materials Engineering, https://doi.org/10.1007/978-3-030-00647-1_6

- controlled rolling with the end of deformation in the lower region of the γ field and subsequent accelerated cooling, which starts from this region and finishes in the bainitic transformation region (CR + AC).

Additionally, the plates subjected to quenching, and tempering (LCR + Q + T) after low-temperature controlled rolling, have been studied.

The centerline segregation zone of the plates after processing by controlled rolling regimes with the end of deformation in the single-phase γ field and in the two-phase ($\gamma + \alpha$) field (HCR and LCR) followed by air cooling exhibits structural heterogeneity in the form of extended lamellar pearlite bands (Fig. 6.1a, b).

The central structural heterogeneity was substantially less pronounced in the plates manufactured by controlled rolling technology with accelerated cooling and quenching with tempering (Fig. 6.1c, d) than in the plates after HCR and LCR. In the axial zone of the plate after CR + AC, the segregation bands with high-carbon bainite regions were present, and almost no structural heterogeneity was detected in the plates after LCR + Q + T.

After all deformation-thermal treatment regimes under study, the microhardness of the axial zone of the plates was higher than that of the base metal (Fig. 6.2). The lowest microhardness was obtained after processing by the HCR scheme. The microhardness of the base metal of the plates after LCR, CR + AC, and LCR + Q + T regimes was at a level of 176–181 $HV_{0.2}$. At the same time, the microhardness of the axial zone was highest after processing by the LCR scheme. The microstructural heterogeneity parameters $\Delta HV_{0.2}$ and $K(HV_{0.2})$ over thickness of the plates manufactured by the CR + AC and LCR + Q + T technologies were lower than those after HCR and LCR. This shows a higher homogeneity of the microstructure over the thickness of the plates after processing by these technological regimes.

Figure 6.3 shows the effect of various deformation-thermal treatment schemes on the HIC resistance of the plates. The HCR and LCR technologies do not provide high HIC resistance of steel. The plates treated by the CR + AC and LCR + Q + T technologies were characterized by high HIC resistance. No hydrogen-induced cracks were found in the tested specimens. The hydrogen-induced cracks detected in the plates after HCR and LCR propagate in the axial plate zone along segregation bands consisting of lamellar pearlite (Fig. 6.4).

Low HIC resistance of the plates produced by HCR and LCR regimes is caused by the formation of the segregation structural heterogeneity in the form of extended lamellar pearlite bands enriched with carbon and manganese in the axial zone, along which the hydrogen-induced cracks propagate. The formation of such bands is facilitated by the diffusion processes occurring in the pearlite region at low cooling rates after controlled rolling. An intense cooling of the plate upon the treatment by the CR + AC and LCR + Q + T regimes suppresses the pearlitic transformation and inhibits the diffusion of carbon. This results in a more uniform ferritic–bainitic structure over cross section of the plate and, as a consequence, in an increased HIC resistance of the steel.

The limitation of carbon and manganese contents, which is necessary to increase the HIC resistance of the plates, reduces the ability to provide the necessary strength

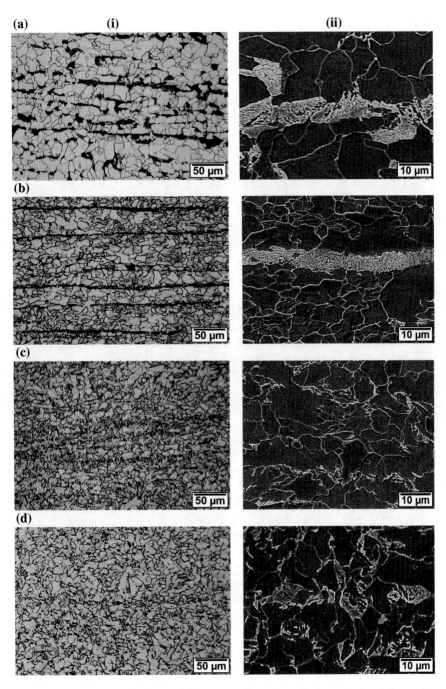

Fig. 6.1 Microstructure of the centerline segregation zone of plates produced by various deformation-thermal treatment schemes, i—OM; ii—SEM: **a**—HCR; **b**—LCR; **c**—CR + AC; **d**—LCR + Q + T

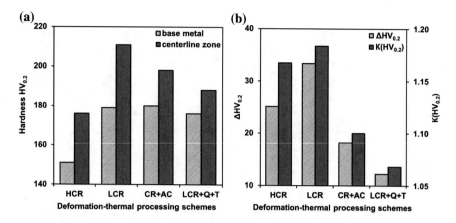

Fig. 6.2 Microhardness of the base metal and centerline segregation zone (**a**) and the $\Delta HV_{0.2}$ and $K(HV_{0.2})$ parameters (**b**) of the plates produced by various deformation-thermal treatment schemes

characteristics of the plates, especially from the steels of X65 and higher grades. To increase the strength properties of the plates manufactured by controlled rolling technology followed by air cooling, it is necessary to use the rolling schemes, at which the metal temperature at the final deformation stage is below the start temperature of the austenite-to-ferrite transformation (Ar_3 temperature). However, controlled rolling with the end of deformation in the two-phase ($\gamma + \alpha$) field promotes an increase in the pearlite banding of the steel structure, an enhancement of segregation structural heterogeneity, an increase in dislocation density, and the flattening of manganese sulfides. All such factors detrimentally affect the HIC resistance of plates.

As the plates are tested in slightly aggressive Solution B, HIC occurs neither after the CR nor after the CR + AC treatment regimes of the plates (Fig. 6.5) (Barthold et al. 1989). At the same time, as the plates are tested in more severe Solution A, a decrease finishing rolling temperature of plates leads to a reduction in their HIC resistance at a simultaneous increase in the ultimate tensile strength of the steel. The plates after CR + AC exhibited a higher HIC resistance.

The negative effect of pearlite banding in both the base metal and the axial zone on the HIC resistance of plates can be reduced by applying tempering after controlled rolling (Usinor Aciers 1987). Tempering above 650 °C allows a partial globulariza-tion of cementite in pearlite and leads to the annihilation of a significant fraction of the dislocations formed at a low finishing rolling temperature. At the same time, high-temperature tempering positively affecting the HIC resistance of plates decreases their strength properties.

The data on the effect of the finishing temperature of the final deformation stage on the HIC resistance of the plates produced from the 09GSF steel in Mill 5000 at Vyksa Steel Works are given in (Barykov 2016). Roughing rolling was performed at temperatures of the completion of austenite recrystallization, and the finishing rolling was performed by three regimes, at which the end deformation temperature

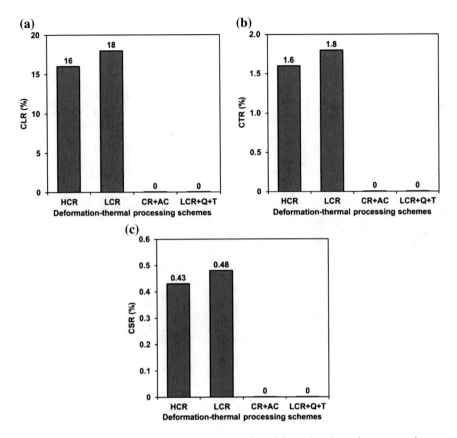

Fig. 6.3 HIC parameters of the plates produced by various deformation-thermal treatment schemes: a—CLR; b—CTR; c—CSR

was $Ar_3 + 50$ °C, $Ar_3 + 30$ °C, or $Ar_3 + 10$ °C. The mechanical properties of the plates produced at different end temperatures above the Ar_3 point were the same, but their HIC resistances were different (Fig. 6.6). The authors showed that the temperature regime of CR with the end of deformation at $Ar_3 + 50$ °C ensures a full absence of hydrogen cracks in the specimens. At a finishing rolling temperature of $Ar_3 + 30$ °C, the CLR parameter was at an acceptable level and did not exceed 6%. The plate rolled at $T_{fr} = Ar_3 + 10$ °C exhibited the worst results, its CLR parameter in some cases reached 24%.

Such effect of the end temperature of the final deformation stage is explained by an increase in the structure anisotropy with decreasing T_{fr}. The austenite structure before the start of cooling after rolling at $T_{fr} = Ar_3 + 50$ °C was characterized by the smallest degree of work hardening and provided the formation of a homogeneous final structure of the plate. With decreasing rolling temperature, the degree of recrystallization decreases, and the microstructure consists mainly of deformed austenite. Cooling of such structure leads to banding in the finished rolled product. In this case,

(a)

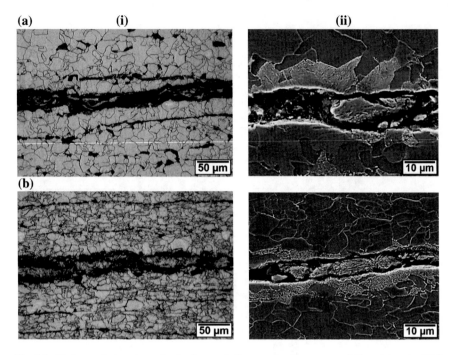

Fig. 6.4 Hydrogen-induced cracks in the centerline segregation zone of the plates, i—OM; ii—SEM: **a**—HCR; **b**—LCR

Fig. 6.5 Effect of the finishing rolling temperature on the parameter of the hydrogen crack length CLR (**a, b**) and the ultimate tensile strength (**c**) of plates 25.4 mm thick from the X52 grade steel: **a**—test in Solution B; **b**—test in Solution A, (white circle—CR; black circle—CR + AC) (Barthold et al. 1989)

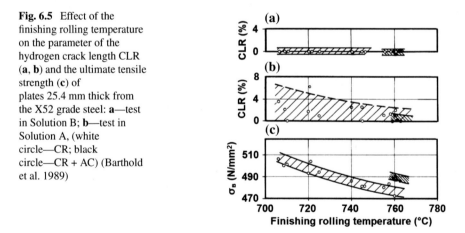

extended planes with a high density of non-metallic inclusions are formed at the boundaries of elongated austenite grains arise.

The technology of quenching after special heating with additional tempering is used for the manufacture of rolled products with strict requirements for HIC resistance, especially in the absence of a unit for accelerated controlled cooling. The

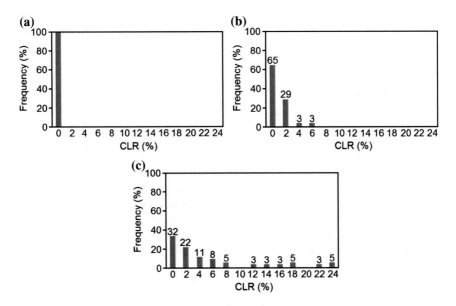

Fig. 6.6 Effect of the finish rolling temperature on the crack length ratio (CLR) of the 09GSF steel plates: **a**—T_{fr} = Ar₃ + 50 °C; **b**—T_{fr} = Ar₃ + 30 °C; **c**—T_{fr} = Ar₃ + 10 °C (Barykov 2016)

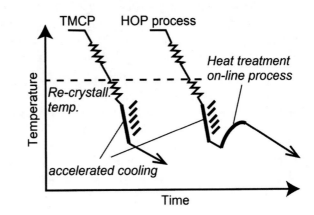

Fig. 6.7 Scheme of plate treatment by the CR + AC technology and HOP (Shabalov et al. 2017)

Russian 13KhFA, 09GSF, 08GBF-U, and 07GBF-U steels were developed for this technology. This provided the required K48-K56 (X52-X65) grade strength level of the plates and their high cracking resistance in hydrogen sulfide-containing media. At present, the technology of controlled rolling with accelerated cooling is also used for the processing of such steels.

A promising technology for efficient control of the processes of structure formation is the use of heat treatment on-line process (HOP) immediately after accelerated cooling (Ishikawa et al. 2009) (Fig. 6.7).

Table 6.1 Experimental CR + AC regimes for the plates 18 mm thick from 0.05%C-1.25%Mn-0.001%S-0.010%P-0.70%(Cr + Ni + Cu)-(Ti + Nb + V)-steel

Treatment regime	T_{fr} (°C)	T_{sc} (°C)	T_{fc} (°C)	V_c (°C/s)
I (T_{fr} > Ar$_3$, T_{sc} > Ar$_3$)	Ar$_3$ + 60	Ar$_3$ + 20	540–560	16–18
II (T_{fr} > Ar$_3$, T_{sc} < Ar$_3$)	Ar$_3$ + 15	Ar$_3$ − 20		
III (T_{fr} < Ar$_3$, T_{sc} < Ar$_3$)	A$_{r3}$ − 20	A$_{r3}$ − 40		

From the point of view of the microstructure formation, HOP allows a structure homogenization in combination with more effective precipitation hardening, which cannot be achieved by conventional thermomechanical processing with accelerated cooling. In this case, the plate after accelerated cooling to a temperature of about 500 °C is reheated to 650 °C with further slow cooling. This prevents the formation of the MA-constituent particles and increases the HIC resistance of plates, while the precipitation of fine carbonitride particles can increase the strength properties.

6.2 Post-deformation Cooling

This section is devoted to the effect of the deformation cooling parameters such as start temperature of accelerated cooling (T_{sc}), finish temperature of accelerated cooling (T_{fc}), and cooling rate (V_c) on the microstructure of the central segregation zone and the HIC resistance of rolled plates.

Start Temperature of Accelerated Cooling The effects of T_{fr} and T_{sc} on the microstructure and properties of plates 18 mm thick from the 0.05%C-1.25%Mn-0.001%S-0.010%P-0.70%(Cr + Ni + Cu)-(Ti + Nb + V)-steel produced by the CR + AC regimes were studied in (Matrosov et al. 2015). The T_{fr} and T_{sc} parameters of the test plates were changed, while the finish temperature of accelerated cooling and the cooling rate were the same (Table 6.1).

Figure 6.8 shows the microstructure of the central segregation zone of the test plates after CR + AC at different start temperatures of accelerated cooling. As T_{fr} and T_{sc} decrease to the two-phase ($\gamma + \alpha$) field, the central structural heterogeneity increases relative to that observed after accelerated cooling from the γ field. High-carbon phases in the segregation zone of the plate treated by regime III (T_{fr} < Ar$_3$, T_{sc} < Ar$_3$) were present in the form of MA constituent and cementite.

The results of microhardness measurement of the base metal and axial zone and the $\Delta HV_{0.2}$ and $K(HV_{0.2})$ parameters of the experimental plates are shown in Fig. 6.9. As the start temperature of accelerated cooling decreases below the Ar$_3$ point, the microhardness of the base metal decreases with a simultaneous increase in microhardness of the central zone of the plate. As a result, $\Delta HV_{0.2}$ increases from 39 to 69 $HV_{0.2}$, and $K(HV_{0.2})$ increases from 1.22 to 1.42.

Figure 6.10 shows the dependences of the HIC parameters on the start temperature of accelerated cooling of plates.

(a)   **(i)**   **(ii)**

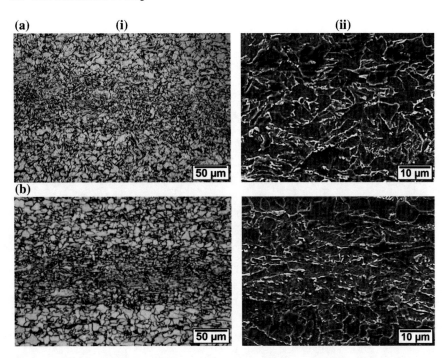

Fig. 6.8 Effect of the start temperature of accelerated cooling on the microstructure of the centerline segregation zone of 0.05%C-1.25%Mn-0.001%S-0.010%P-0.70%(Cr + Ni + Cu)-(Ti + Nb + V)-steel plates, i—OM; ii—SEM: **a**—$T_{fr} > Ar_3$, $T_{sc} > Ar_3$; **b**—$T_{fr} < Ar_3$, $T_{sc} < Ar_3$

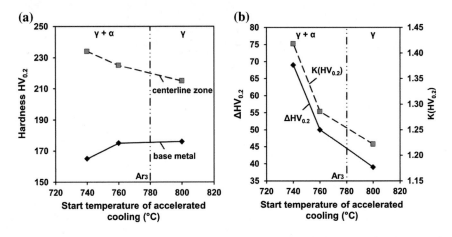

Fig. 6.9 Effect of start temperature of accelerated cooling on the microhardness of the base metal and the centerline segregation zone (**a**) and the $\Delta HV_{0.2}$ and $K(HV_{0.2})$ parameters (**b**) of the 0.05%C-1.25%Mn-0.001%S-0.010%P-0.70%(Cr + Ni + Cu)-(Ti + Nb + V)-steel plates

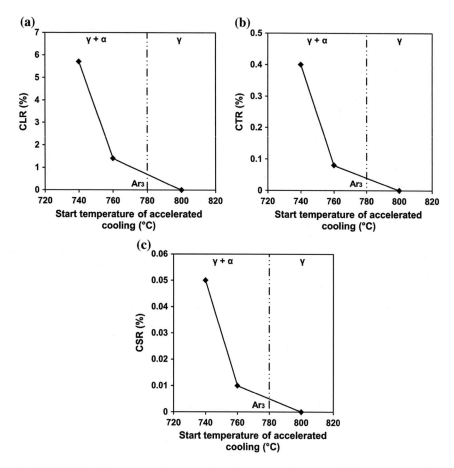

Fig. 6.10 Effect of the start temperature of accelerated cooling on the HIC parameters of the 0.05%C-1.25%Mn-0.001%S-0.010%P-0.70%(Cr + Ni + Cu)-(Ti + Nb + V)-steel plates: **a**—CLR; **b**—CTR; **c**—CSR

If the start temperature of accelerated cooling belongs to the γ field, the HIC parameters tend to zero. On the contrary, as T_{sc} decreases to the $(\gamma + \alpha)$ field, and the HIC parameters increase. The detected hydrogen-induced cracks are present in the axial rolling zone (Fig. 6.11). The crack propagation occurs along the interphase interfaces between ferrite and hard high-carbon structures, mainly the MA constituent.

On the basis of the experimental results, one can conclude that, as the start temperature of accelerated cooling is above the Ar_3 temperature, the microstructure is more uniform over the plate thickness and less susceptible to HIC than that formed upon accelerated cooling from the temperature below Ar_3. This is explained by the enrichment of undecomposed austenite with carbon and alloying elements upon the

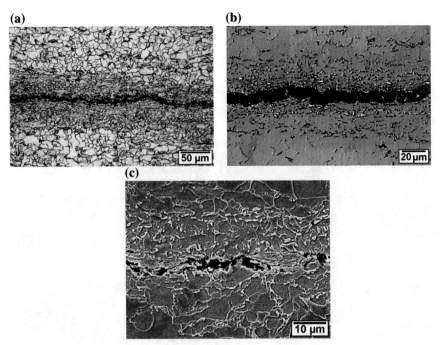

Fig. 6.11 HIC in the centerline segregation zone of the 0.05%C-1.25%Mn-0.001%S-0.010%P-0.70%(Cr + Ni + Cu)-(Ti + Nb + V)-steel plate treated at $T_{fr} <$ Ar$_3$ and $T_{sc} <$ Ar$_3$: **a**—OM; **b**—OM (etching in LePera reagent); **c**—SEM

$\gamma \rightarrow \alpha$ polymorphic transformation occurring upon slow cooling. The decomposition of such austenite can occur in the pearlitic and bainitic regions, as well as with the formation of the MA constituent. In this case, the difference between the steel matrix and the segregation bands in hardness increases, which promotes the nucleation and propagation of hydrogen-induced cracks.

Finish Temperature of Accelerated Cooling The effect of the finish temperature of accelerated cooling on the microstructure of the central segregation zone and the degree of its heterogeneity over the plate thickness is shown in (Matrosov et al. 2015) by the example of industrial plates 20 mm thick from the 0.07%C-1.33%Mn-0.001%S-0.010%P-0.70%(Cr + Ni + Cu)-0.120%(Ti + Nb + V)-steel. The test plates were cooled from the austenite field to temperatures in a range of 430–610 °C at rates of 15–20 °C/s.

The microstructure of the central segregation zone of plates accelerated cooled to temperatures of 610, 500, and 430 °C is shown in Fig. 6.12. The axial zone of the rolled plates after accelerated cooling to 610 °C exhibits a substantial segregation structural heterogeneity in the form of degenerate pearlite bands.

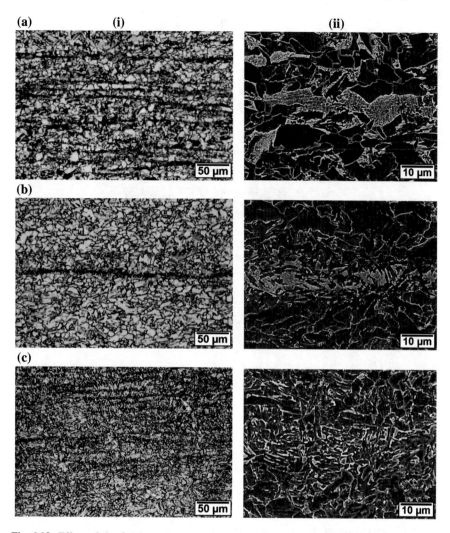

Fig. 6.12 Effect of the finish temperature of accelerated cooling on the microstructure of the centerline segregation zone of the 0.07%C-1.33%Mn-0.001%S-0.010%P-0.70%(Cr + Ni + Cu)-0.120%(Ti + Nb + V)-steel plates, i—OM; ii—SEM: **a**—$T_{fc} = 610$ °C; **b**—$T_{fc} = 500$ °C; **c**—$T_{fc} = 430$ °C

After accelerated cooling to $T_{fc} = 500$ °C, the segregation zone is much less pronounced and consists of bands with high-carbon bainite regions.

The microstructure of the centerline segregation zone after cooling to 430 °C exhibits numerous bands consisting of the MA-constituent regions clearly visible after etching in the LePera reagent (Fig. 6.13).

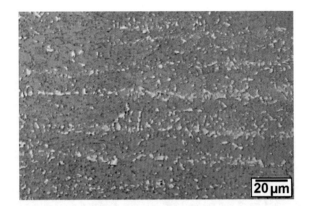

Fig. 6.13 Microstructure of the centerline segregation zone of the 0.07%C-1.33%Mn-0.001%S-0.010%P-0.70%(Cr + Ni + Cu)-0.120%(Ti + Nb + V)-steel plate after accelerated cooling to T_{fc} = 430 °C, OM (etching in a LePera reagent)

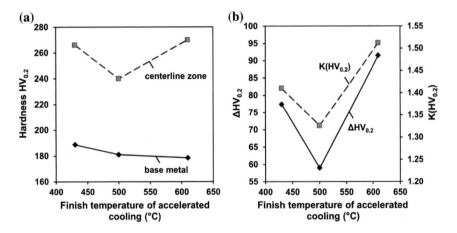

Fig. 6.14 Effect of finish temperature of accelerated cooling on the microhardness of the base metal and the centerline segregation zone (**a**) and the $\Delta HV_{0.2}$ and $K(HV_{0.2})$ parameters (**b**) of the 0.07%C-1.33%Mn-0.001%S-0.010%P-0.70%(Cr + Ni + Cu)-0.120%(Ti + Nb + V)-steel plates

The microhardness and the degree of microstructural heterogeneity over the plate thickness are shown in Fig. 6.14. As the finish temperature of accelerated cooling decreases from 610 to 430 °C, the microhardness of the base metal of the plates increases comparatively weakly, remaining within 180–190 $HV_{0.2}$. The microhardness of the axial zone for all T_{fc} was substantially higher than that of the base metal. As T_{fc} decreases from 610 to 500 °C, the microhardness of the axial zone substantially decreases, from 270 $HV_{0.2}$ to 240 $HV_{0.2}$. A further decrease in T_{fc} to 430 °C leads to an increase in the microhardness of the axial zone to 266 $HV_{0.2}$. The parame-

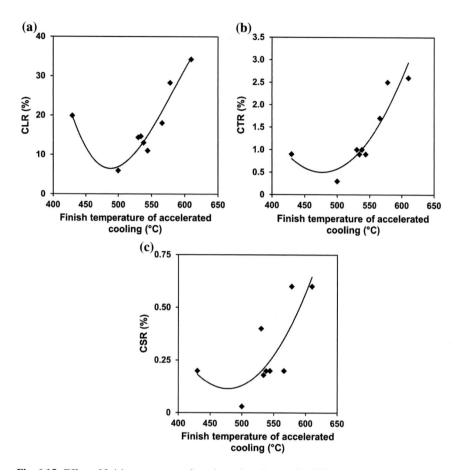

Fig. 6.15 Effect of finish temperature of accelerated cooling on the HIC parameters of the 0.07%C-1.33%Mn-0.001%S-0.010%P-0.70%(Cr + Ni + Cu)-0.120%(Ti + Nb + V)-steel plates: **a**—CLR; **b**—CTR; **c**—CSR

ters of microstructural heterogeneity over the plate thickness, $\Delta HV_{0.2}$ and $K(HV_{0.2})$, change correspondingly to the microhardness of the axial zone and are smallest at $T_{fc} = 500\ °C$. At the same time, the difference in microhardnesses of the base metal and axial zone remains as high as ~60 $HV_{0.2}$.

The effect of the finish temperature of accelerated cooling on the HIC resistance of plates is associated with the microstructure type (Fig. 6.15).

For all tested finish temperatures of accelerated cooling, the detected hydrogen-induced cracks propagate over the structure of segregation bands in the centerline segregation zone. Such structures consist of degenerate pearlite in the plates cooled to 610 °C (Fig. 6.16a) and of high-carbon bainite in the plates cooled to 500 °C

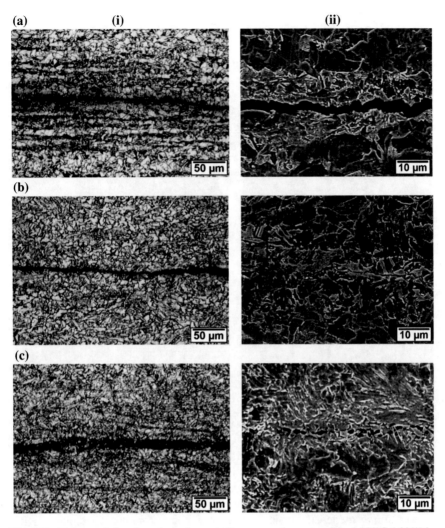

Fig. 6.16 Hydrogen-induced cracks in the centerline segregation zone of the 0.07%C-1.33%Mn-0.001%S-0.010%P-0.70%(Cr + Ni + Cu)-0.120%(Ti + Nb + V)-steel plates, i—OM; ii—SEM: **a**—T_{fc} = 610 °C; **b**—T_{fc} = 500 °C; **c**—T_{fc} = 430 °C

(Fig. 6.16b). After cooling to 430 °C, the cracks propagate along the boundaries between the MA constituent and ferrite (Figs. 6.16c and 6.17).

Similar dependences of the properties of rolled steels differing in alloying and microalloying systems and contents of carbon and manganese on the finish temperature of accelerated cooling were obtained in (Tamehiro et al. 1985) (Table 6.2). The graphs demonstrated in Fig. 6.18 show that, for all four investigated steels, the optimal combination of mechanical properties and cold resistance, as well as the min-

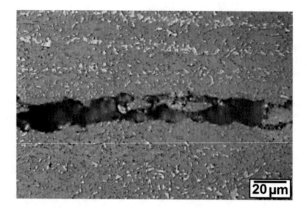

Fig. 6.17 HIC in the centerline segregation zone of the 0.07%C-1.33%Mn-0.001%S-0.010%P-0.70%(Cr + Ni + Cu)-0.120%(Ti + Nb + V)-steel plate after accelerated cooling to 430 °C, OM (etching in LePera reagent)

Table 6.2 Chemical composition of the steels of various alloying systems (Tamehiro et al. 1985)

Steel no.	Elements (wt%)									C_{eq}
	C	Si	Mn	P	S	Nb	V	Mo	Others	
1	0.104	0.30	1.05	0.004	0.0010	0.04	–	–	Ni, Cu, Ti, Ca	0.31
2	0.082	0.28	1.33	0.007	0.0009	0.04	0.09	–	Ni, Ti, Ca	0.34
3	0.057	0.17	1.08	0.005	0.0013	0.04	–	0.25	Ni, Ti, Ca	0.31
4	0.024	0.19	1.17	0.009	0.0017	0.04	–	–	Ni, Cu, Ti, Ca	0.25

imum CAR upon HIC, is provided as the finish temperature of accelerated cooling lies in a temperature range of 450–550 °C.

The microstructure with lamellar pearlite is obtained in the cases where the accelerated cooling is not applied, or is finished at too high temperature, or cooling is performed at too low rate. Hydrogen is accumulated at the ferrite–pearlite interfaces, and this leads to HIC. A lower finish temperature of accelerated cooling can lead to the formation of martensite, which is also undesirable from the standpoint of HIC resistance. In a range of the finish temperature of accelerated cooling of 450–550 °C, the ferritic–bainitic microstructure is more homogeneous and less susceptible to HIC.

Thus, the results of the study show that, in the case of relatively high finish temperatures of accelerated cooling, the structure formed in the axial zone of the plates contains a second phase in the form of coarse pearlite bands. Accelerated cooling to temperatures belonging to the lower range of bainitic transformation leads to the formation of bands consisting of high-carbon bainite. At a lower finish temperature of accelerated cooling, bands with regions of the MA constituent can form. Such high-carbon structures are favorable for the nucleation and propagation of hydrogen-induced cracks. The optimum range of the finish temperatures of accelerated cooling for the steels under study is the lower range of the bainitic transformation region

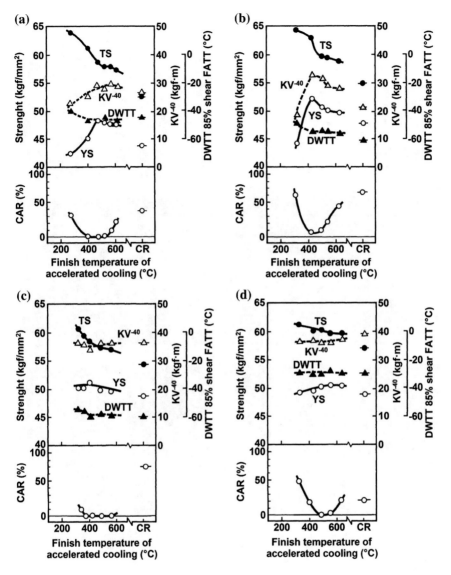

Fig. 6.18 Effect of the finish temperature of accelerated cooling on the mechanical properties and the area of the hydrogen crack (CAR): **a**—steel No. 1; **b**—steel No. 2; **c**—steel No. 3; **d**—steel No. 4 (Tamehiro et al. 1985)

(in this case, $T_{fc} \sim 500$ °C). The ferritic–bainitic microstructure formed in the axial zone at such temperatures is more uniform and less susceptible to HIC than the structures formed upon accelerated cooling to higher or lower temperatures.

(a)   **(i)**   **(ii)**

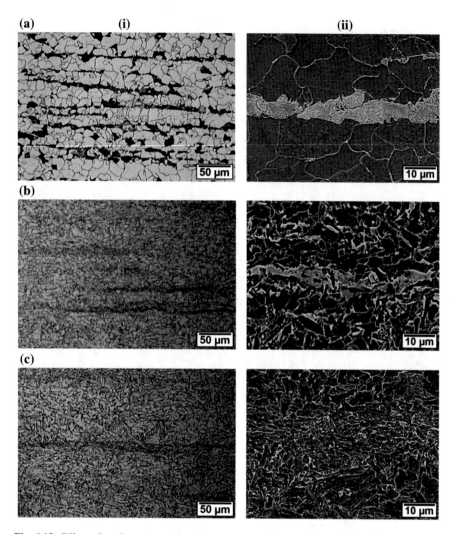

Fig. 6.19 Effect of cooling rate on the microstructure of the centerline segregation zone of the 0.05%C-1.22%Mn-0.001%S-0.010%P-0.65%(Cr + Ni + Cu)-0.10%(Ti + Nb + V)-steel plates, i—OM; ii—SEM: **a**—V_c = 2 °C/s; **b**—V_c = 15 °C/s; **c**—V_c = 25 °C/s

Cooling Rate The effect of the cooling rate on the microstructure and properties was studied on the 0.05%C-1.22%Mn-0.001%S-0.010%P-0.65%(Cr + Ni + Cu)-0.10%(Ti + Nb + V)-steel plates 20 mm thick after cooling at rates of 2, 15, and 25 °C/s. The start temperature of accelerated cooling of the plates was Ar_3 + (20–30) °C, and the finish temperature was in a range of 520–540 °C (Matrosov et al. 2015).

The microstructure of the central segregation heterogeneity zones of the experimental plates is shown in Fig. 6.19.

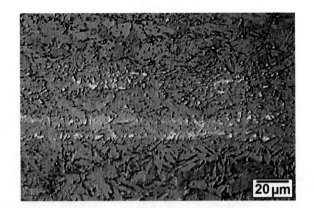

Fig. 6.20 Microstructure of the centerline segregation zone of the 0.05%C-1.22%Mn-0.001%S-0.010%P-0.65%(Cr + Ni + Cu)-0.10%(Ti + Nb + V)-steel plate cooled at $V_c = 25$ °C/s, OM (etching in the LePera reagent)

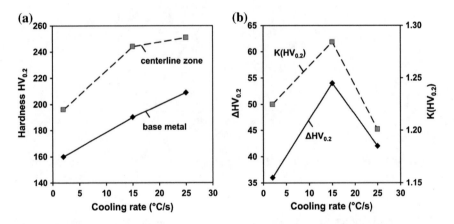

Fig. 6.21 Effect of the cooling rate on the microhardness of the base metal and centerline segregation zone (**a**) and the $\Delta HV_{0.2}$ and $K(HV_{0.2})$ parameters (**b**) of the 0.05%C-1.22%Mn-0.001%S-0.010%P-0.65%(Cr + Ni + Cu)-0.10%(Ti + Nb + V)-steel plates

Upon cooling of the plate at $V_c = 2$ °C/s, numerous segregation bands consisting of lamellar pearlite are formed in the axial zone of the plate. The segregation bands in the axial zone of the plate cooled at a rate of 15 °C/s consist of extended regions of high-carbon upper bainite. As the cooling rate increases to 25 °C/s, particles of the MA constituent are observed in the microstructure of the segregation bands (Figs. 6.19c and 6.20). At the same time, the central structural heterogeneity is less pronounced, and the segregation bands are less extended than those are in microstructure of the axial zone of the plates cooled at rates of 2 and 15 °C/s.

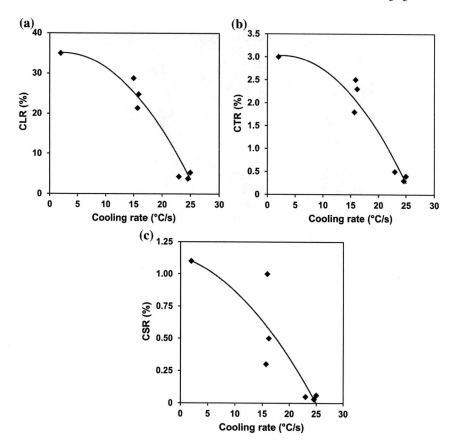

Fig. 6.22 Effect of the cooling rate on the HIC parameters of the 0.05%C-1.22%Mn-0.001%S-0.010%P-0.65%(Cr + Ni + Cu)-0.10%(Ti + Nb + V)-steel plates: **a**—CLR; **b**—CTR; **c**—CSR

Figure 6.21 shows the effect of the cooling rate on the microhardness of the base metal and the centerline segregation zone and on the $\Delta HV_{0.2}$ and $K(HV_{0.2})$ parameters of the test plates. With increasing cooling rate, the microhardness of both the base metal and the axial zone of the plates increases. Low $\Delta HV_{0.2}$ and $K(HV_{0.2})$ values of the plate cooled at $V_c = 2$ °C/s are due to the formation of pearlite banding in the axial zone. The structural homogeneity of the plates after cooling at $V_c = 25$ °C/s is higher than that is at $V_c = 15$ °C/s.

Figure 6.22 shows that, with an increase in the cooling rate from 2 to 25 °C/s, a significant increase in the HIC resistance of the plates is observed, which is expressed in a significant decrease in the CLR, CTR, and CSR parameters.

All hydrogen-induced cracks propagate along the segregation bands in the axial zone consisting of high-carbon structures (Fig. 6.23). In the plate cooled in air, cracking occurs along the bands of lamellar pearlite. In the plate cooled at a rate of 15 °C/s, the hydrogen-induced cracks propagate along high-carbon bainite bands

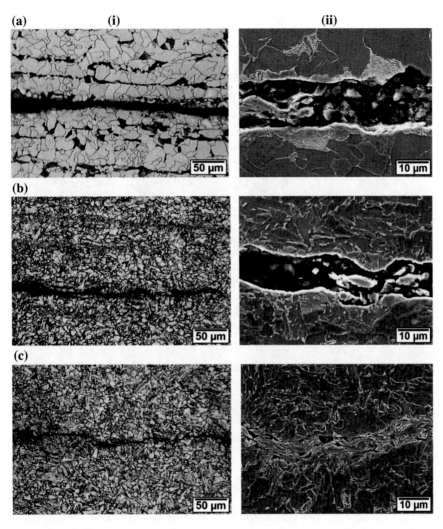

Fig. 6.23 Hydrogen-induced cracks in the centerline segregation zone of the 0.05%C-1.22%Mn-0.001%S-0.010%P-0.65%(Cr + Ni + Cu)-0.10%(Ti + Nb + V)-steel plates after different V_c, i—OM; ii—SEM: **a**—$V_c = 2$ °C/s; **b**—$V_c = 15$ °C/s; **c**—$V_c = 25$ °C/s

contained in the centerline segregation zone. HIC in the axial zone of a plate cooled at a rate of 25 °C/s occurs along the interfaces between ferrite and MA-constituent particles (Figs. 6.23c and 6.24).

An increase in the microstructural homogeneity over the plate thickness and an increase in the HIC resistance of the plates with increasing cooling rate are associated with a decrease in the rate of the diffusion of carbon and other elements into austenite upon phase transformations. This allows one to reduce the amount of non-decomposed austenite regions enriched with alloying elements, primarily with car-

Fig. 6.24 Hydrogen-induced crack in the centerline segregation zone of the 0.05%C-1.22%Mn-0.001%S-0.010%P-0.65%(Cr + Ni + Cu)-0.10%(Ti + Nb + V)-steel plate cooled at $V_c =$ 25 °C/s, OM (etching in a LePera reagent)

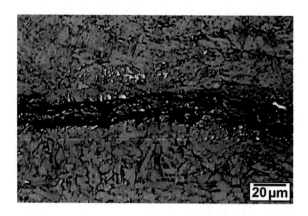

bon, and to reduce the number and extension of the bands consisting of unfavorable transformation products such as pearlite, high-carbon bainite, and MA constituent, which adversely affect the HIC resistance.

References

Barthold, G., Streißelberger, A., Bauer, J. (1989). Modern line pipe steels for sour service—Experience in applying TM-rolled and accelerated cooling plate. *Technical information Dillinger Huttenwerke, 9.*

Barykov, A. B. (Ed.). (2016). *Development of steel manufacturing technology for steel rolled product and pipe in the Vyksa Production Area: Coll. Works.* Moscow: Metallurgizdat.

Ishikawa, N., et al. (2009). Recent advance in high strength linepipe for heavy sour service. In: *Pipe Technology Conference*, Ostend, Belgium, 12–14 October 2009.

Kuznechenko, Y. S., Shabalov, I. P., Kholodnyy, A. A., et al. (2017). Centerline segregation inhomogeneity and resistance to hydrogen induced cracking of rolled plates from pipe steels. Part 2. Influence of thermo-mechanical treatment. *Problems of Ferrous Metallurgy and Materials Science, 3,* 85–94.

Matrosov, Y. I., Kholodnyi, A. A., Matrosov, M. Y., et al. (2015). Effect of accelerated cooling parameters on microstructure and hydrogen cracking resistance of low-alloy pipe steels. *Metallurgist, 59*(1), 60–68.

Matrosov, Y. I, Kholodnyi, A. A., Popov, E. S., et al. (2014). Influence of thermomechanical processing and heat treatment on microstructure formation and HIC resistance of pipe steel. *Problems of Ferrous Metallurgy and Materials Science, 1,* 98–104.

Shabalov, I. P., Matrosov, Y. I, Kholodnyi, A. A., et al. (2017). *Steel for gas and oil pipelines resistant to fracture in hydrogen sulphide-containing media.* Moscow: Metallurgizdat.

Tamehiro, H., et al. (1985). Effect of accelerated cooling after controlled rolling on the hydrogen induced cracking resistance of line pipe steel. *Transactions of the Iron and Steel Institute of Japan, 25,* 982–988.

Usinor Aciers. (1987*). Steel grades for the manufacture of welded pipes, resistant to cracking under the influence of hydrogen sulphide.* Document of the Working Group on the Metallurgical Industry of Franco-Soviet Cooperation, p. 51.

Chapter 7
Manufacturing Technology of Steels for Pipes Ordered for Sour Service

In this chapter, the state of the technology of industrial production of H_2S-resistant rolled plates and strips and pipe manufacture by leading Russian and international metallurgical and pipe enterprises is considered. At present, mainly pipes of grades up to X65 inclusive have found commercial use for the construction of sour-gas-resistant pipelines. Research works aimed at the development of X70 and X80 grade pipes for operation in the most aggressive media are underway. To achieve high properties of pipe steels, including HIC and SSC resistances, high strength, cold resistance, weldability, it is necessary to choose the optimal chemical composition of steel and to optimize the production methods such as steelmaking, hot plastic deformation, and cooling. An important condition is the formation of fine structure with the regulated number and size of structure constituents (ferrite, pearlite, bainite), fine carbonitride particles precipitated under certain rolling conditions, and limited quantity of non-metallic inclusions.

The results of manufacturing large batches of K48 (X52)–K52 (X60) grades rolled plates and large diameter electric-welded pipes at the Heavy Plate Mill 5000 and in the Electrical Resistance Welding Shop of the Vyksa Steel Works are presented in (Barykov 2016). In 2014–2015, large batches of K48 (X52) grade plates 13.0–16.0 mm thick and K50 (X56) grade plates 26.0–28.0 mm thick of more than 20,000 tons in total were produced. The plates were made from slabs produced by the Novolipetsk Iron and Steel Works and characterized by a high quality of macrostructure and a low content of harmful impurities (S $\leq$ 0.002%, P $\leq$ 0.010%) and gases (N $\leq$ 0.0045%, H $\leq$ 0.0002%, O $\leq$ 0.0025%). The plates and the pipes made of them fully meet the requirements for mechanical properties and cold resistance (Table 7.1) and correspond to the C-2 level with respect to HIC resistance and sulfide stress cracking.

The production of plates 29.0 mm thick from the K52 (X60) grade steel was also tested. The chemical composition and mechanical properties of the steel are shown in Tables 7.2 and 7.3. The corrosion tests showed a high HIC resistance of the plates. No hydrogen-induced cracks were found in the specimens of the plates after the tests (Barykov 2016).

© Springer Nature Switzerland AG 2019
I. Shabalov et al., *Pipeline Steels for Sour Service*, Topics in Mining, Metallurgy and Materials Engineering, https://doi.org/10.1007/978-3-030-00647-1_7

Table 7.1 Mechanical properties of plates and base metal of pipes of 13.0–16.0 mm in plate/wall thickness for the K48 (X52) grade and 26.0–28.0 mm in plate/wall thickness for the K50 (X56) grade, (min–max)/average (Barykov 2016)

Grade	Plate/pipe	$\sigma_{0.5}$ (N/mm²)	σ_B (N/mm²)	δ_5 (%)	$\sigma_{0.5}/\sigma_B$	HV_{10}	KCV^{-20} (J/cm²)	KCU^{-60} (J/cm²)	DWTT^{-20} (%)
K48 (X52)	Plate	$\dfrac{380–490}{410}$	$\dfrac{470–520}{500}$	$\dfrac{26–37}{27}$	$\dfrac{0.84–0.89}{0.87}$	$\dfrac{149–168}{154}$	$\dfrac{430–470}{450}$	$\dfrac{410–450}{430}$	100
	Pipe	$\dfrac{375–440}{409}$	$\dfrac{475–520}{500}$	$\dfrac{25–35}{24}$	$\dfrac{0.76–0.86}{0.82}$	$\dfrac{150–170}{160}$	$\dfrac{380–440}{410}$	$\dfrac{350–440}{405}$	$\dfrac{90–100}{}$
K50 (X56)	Plate	$\dfrac{420–520}{470}$	$\dfrac{500–580}{540}$	$\dfrac{23–31}{28}$	$\dfrac{0.83–0.89}{0.87}$	$\dfrac{160–200}{180}$	$\dfrac{300–440}{380}$	$\dfrac{310–460}{390}$	$\dfrac{70–100}{90}$
	Pipe	$\dfrac{435–500}{470}$	$\dfrac{520–580}{550}$	$\dfrac{24–33}{27}$	$\dfrac{0.81–0.88}{0.85}$	$\dfrac{170–210}{190}$	$\dfrac{380–440}{410}$	$\dfrac{360–440}{400}$	$\dfrac{70–100}{85}$

Table 7.2 Chemical composition of the K52 (X60) grade steel (Barykov 2016)

Elements (wt%)					
C	Mn	Si	S	P	Others
0.04–0.06	1.00–1.10	0.30–0.40	0.001	0.005–0.010	Cu, Ni, Ti, Nb, V

Table 7.3 Mechanical properties of the K52 (X60) grade steel plate 29.0 mm thick (Barykov 2016)

σ_T (N/mm^2)	σ_B (N/mm^2)	δ_5 (%)	$\sigma_{0.5}/\sigma_B$	KCV^{-20} (J/cm^2)	KCV^{-40} (J/cm^2)	DWTT^{-20} (%)
520–550	600–630	25–28	0.85–0.86	390–420	370–390	70–80

The technologies of steelmaking, out-of-furnace processing, continuous casting of steel, and rolling of continuously cast slabs were developed and implemented for the conditions of the Casting and Rolling Complex of the JSC "OMK-Steel" branch. Such technologies provided high HIC resistance with simultaneous achievement of the specified level of strength characteristics and cold resistance of skelp and plates 6.0–12.7 mm thick for the manufacture of up to X60 grade electric-welded pipes of 159–530 mm in diameter (Kudashov et al. 2017).

Steelmaking process provided a low (max 0.06%) carbon content, and the phosphorus content was minimized. Deep deoxidation to a dissolved oxygen content of maximum 5 ppm and deep desulfurization ([S] < 0.0010%) of the steel were performed upon steel processing in the ladle furnace unit. Due to the slag regime optimization and efficient modifying treatment with calcium only or calcium and REM, high purity of steel relative to non-metallic inclusions was obtained. For the removal of hydrogen and nitrogen, the metal was degassed by vacuum treatment. At the stage of casting in CCM, the conditions for uniform heat removal were provided to maintain a uniform surface temperature of the slab and to optimize the secondary cooling regimes ensuring minimum axial segregation and at the same time high surface quality of the slab. At the stage of the slab solidification, "soft reduction" was used. The macrostructure of the slabs manufactured using the developed technology was characterized by low porosity and low chemical heterogeneity of the axial zone, the absence of cracks, and a low level of contamination by sulfide and carbonitride inclusions (Kudashov et al. 2017).

Strip was produced at the 1950 Rolling Mill using thermomechanical processing with accelerated cooling, which led to the formation of a fine structure providing a high HIC resistance and a good cold resistance.

The results of joint research of the effect of chemical composition, controlled rolling regimes, and the temperature rate parameters of post-deformation cooling on the microstructure of the base metal and the central segregation zone, mechanical properties, and HIC resistance of the plates from low-alloy pipe steels in the industrial conditions of the Azovstal iron and steel works and I.P. Bardin TsNIICherMet are presented in Shabalov et al. (2017); Matrosov et al. (2015); Kholodnyi et al. (2016a); Kholodnyi et al. (2017). The data on the production of plates 20.0 mm thick for the manufacture of the X52MS, X56MS, X60MS, and X65MS grade large diameter electric-welded pipes ordered for sour service are considered in Kholodnyi et al. (2016b). The requirements imposed on the chemical composition and mechanical properties of the X52MS-X65MS grade plates are given in Tables 7.4 and 7.5.

The requirements specified for the chemical composition of steels include decreased contents of carbon, manganese, and harmful impurities (S ≤ 0.002%, P ≤ 0.020%). With increasing grade, the permissible content of molybdenum and microalloying elements (Ti, Nb, V) increases. The Charpy impact energy and the shear area of DWTT fracture surface were tested at a temperature of −20 °C.

The following requirements for HIC and SSC resistance in Solution A were imposed on the X52MS-X65MS grade plates and pipes:

- HIC: CLR ≤ 15%, CTR ≤ 5%, CSR ≤ 2%;
- SSC threshold stress for four-point bending test: $\sigma_{thr} \geq 72\%$ of the specified minimum yield strength.

The chemical composition (Table 7.6), the quality requirements for the continuously cast billet, and the regimes of controlled rolling with accelerated cooling for the production of the X52MS, X56MS, X60MS, and X65MS grade plates 20.0 mm thick are recommended on the basis of the research results.

For the reduction of the central segregation heterogeneity of the continuously cast slabs and plates, the heats contained C ≤ 0.07% and Mn ≤ 1.00%, and the mass fraction of harmful impurities was limited to the level S ≤ 0.001% and P ≤ 0.010%. For effective sulfide globularization, calcium treatment was carried out at a Ca/S ratio of ~2. All experimental heats were alloyed with Cu + Ni + Cr ≤ 0.80%, and the heats from the X60MS and X65MS grade steels additionally contained Mo. The steels were microalloyed with complex additions of Ti + Nb + V ≤ 0.120%.

To provide an increased HIC resistance of plates, the segregation heterogeneity of continuously cast slabs should not exceed a rating of 2.0 according to the evaluation of macrotemplates by the Mannesmann scale. For this reason, the steel upon continuous casting was overheated in the intermediate ladle above the liquidus temperature by maximum 25 °C. At the same time, the roller guide of the CCM was adjusted before casting. The use of the optimal chemical composition and continuous casting technology ensured the segregation heterogeneity of slabs at a level of 1.5–2.0.

Thermomechanical processing of plates in the two-stand plate mill 3600 was carried out using technology of controlled rolling followed by accelerated cooling. Slabs 270 mm thick were heated for rolling to temperatures ≤1200 °C for 4–6 h.

Rolling in the roughing stand was carried out at temperatures of the lower temperature range of complete austenite recrystallization with a finish rolling temperature above 980 °C.

At the stage of finishing rolling, deformation was performed in the austenite non-recrystallization temperature range for the formation of an increased density of lattice defects and a larger number of nucleation sites for the new phase. The start rolling temperature in the finishing stand was in a range of 880–920 °C. The finish rolling temperature was specified on the basis of the need to ensure the start temperature of accelerated cooling.

Table 7.4 Requirements for the chemical composition of X52MS-X65MS grade steels

Grade	Elements (wt%), maximum														
	C	Mn	Si	S	P	Ti	Nb	V	Cu	Ni	Cr	Mo	N	Al$_{tot.}$	P$_{cm}$
X52MS	0.10	1.45	0.45	0.002	0.020	0.040	0.060	0.050	0.35	0.30	0.30	0.15	0.012	0.060	0.20
X56MS		1.45				0.040	0.080	0.060				0.15			0.21
X60MS		1.45				0.060	0.080	0.080				0.35			0.21
X65MS		1.60				0.060	0.080	0.100				0.35			0.22

Table 7.5 Requirements for the mechanical properties of the X52MS-X65MS grade plates 20.0 mm thick (transverse direction)

Grade	$\sigma_{0.5}$ (N/mm^2)	σ_B (N/mm^2)	$\delta_{2''}$ (%)	$\sigma_{0.5}/\sigma_B$	KV^{-20} (J)	$DWTT^{-20}$ (%)
X52MS	370–510	460–640	≥30	≤0.90	≥100	≥90
X56MS	400–530	490–670	≥30			
X60MS	425–550	520–700	≥29			
X65MS	460–580	535–740	≥29			

The start temperature of the accelerated cooling was above the phase transformation start temperature (Ar_3), which was determined from the thermokinetic diagrams of the decomposition of hot deformed austenite at a cooling rate of 2 °C/s. This rate is close to the cooling rate of the rolled plate in still air during its transportation from the finishing stand to the accelerated cooling unit. The finish temperature of accelerated cooling was specified to be 520 ± 30 °C for the X52MS, X56MS, and X60MS grade steel plates and 430 ± 30 °C for the X65MS grade steel plates. The accelerated cooling rate was 25 ± 5 °C/s.

After CR + AC, plates were subjected to maturing in stacks in the area of anti-flake treatment to improve the continuity and increase the plasticity and ductility. The stacking temperature was at least 350 °C, the removal temperature was ≤ 100 °C, and the holding time in stacks was ≥48 h.

Figure 7.1 shows the microstructure over the thickness of the X65MS grade steel plates. The plates have a fine-grained ferritic–bainitic microstructure (quasipolygonal ferrite and high-carbon bainite) with a slight central structural heterogeneity. The plate metal was characterized by high cleanliness from non-metallic inclusions.

The mechanical properties of the plates met the specified requirements (Table 7.7). Relative elongation $\delta_{2''}$ and the Charpy impact energy substantially exceeded the specified requirements. Figure 7.2 shows the results of serial tests on impact bending of the plates from the test steels at temperatures ranging from −20 to −80 °C. The fraction of the ductile constituent in the fracture surface of the DWTT specimens was at least 90%.

All plates exhibited a high HIC resistance (Table 7.8) and met the requirements for SSC resistance in a hydrogen sulfide-containing medium.

The experience of the production of skelp 6 mm thick for the electric-welded (HFC) gas–oil and tubing pipes of increased cold resistance and corrosion resistance in the Cherepovets Steel Mill of Severstal is considered in Zikeev et al. (2008), Golovanov et al. (2005). The chemical composition of the steel is shown in Table 7.9.

The qualitative macrostructure characteristics of the continuously cast slabs were sufficiently high (estimated by OST 14-2-72): Axial looseness and segregation bands corresponded to a number of 1.0, the axial chemical inhomogeneity number was 0.5, and the number of axial cracks, nest-like cracks, and point imperfection was 0.

Slabs 250 mm thick were rolled in a continuous wide-strip rolling mill 2000 into strips of 6 × 1050 mm in cross section as coils of 17–19 t in weight. The

Table 7.6 Chemical composition of the X52MS-X65MS grade steels

Grade	Elements (wt%)								
	C	Mn	Si	S	P	Ti + V+Nb	Others	N	Al_tot.
X52MS, X56MS	≤0.07	≤1.00	0.15–0.25	≤0.001	≤0.010	≤0.120	Cu + Ni + Cr ≤ 0.80	≤0.010	0.020–0.040
X60MS, X65MS							Cu + Ni + Cr + Mo ≤ 1.00		

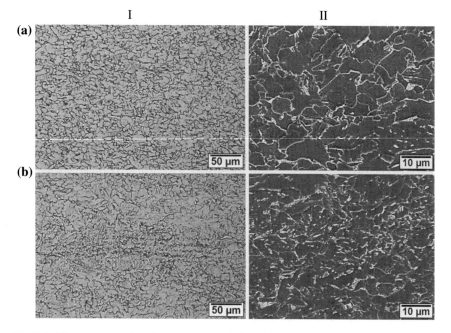

Fig. 7.1 Microstructure of the base metal (**a**) and the axial zone (**b**) of plates 20 mm thick from the X65MS grade steel. I—OM; II—SEM

Table 7.7 Mechanical properties of the X52MS-X65MS grade steel plates 20.0 mm thick. (min—max)/average (Kholodnyi et al. 2016b)

Grade	$\sigma_{0.5}$ (N/mm^2)	σ_B (N/mm^2)	$\delta_{2''}$ (%)	$\sigma_{0.5}/\sigma_B$	KV^{-20} (J)	DWTT^{-20} (%)
X52MS	429–460 443	515–550 532	48–58 52.7	0.80–0.86 0.83	282–348 320	90–100 96.5
X56MS	433–462 450	526–566 543	49–59 51.6	0.80–0.86 0.83	290–356 323	90–95 93.7
X60MS	446–488 465	544–594 564	46–53 48.9	0.81–0.84 0.82	301–398 338	95–100 99.4
X65MS	466–498 482	556–606 585	45–52 48.6	0.79–0.85 0.82	299–354 330	90–95 93.7

heating temperature for rolling was 1240 °C. The thickness of the rolled stock for the finishing group of stands was 36 mm to provide a total reduction by a factor of six. The finish rolling temperature was 850 ± 10 °C, and the finish coiling temperature was 560 ± 10 °C.

The microstructure of the steel was ferritic–pearlitic with a grain size number of 10–11 and a banding number of 0–1. The contents of non-metallic inclusions in the plates corresponded to the following numbers (according to GOST 1778): 0–1.3 for sulfides; 0–1.7 for oxides; and 0.2–1.8 for silicates. The mechanical properties of

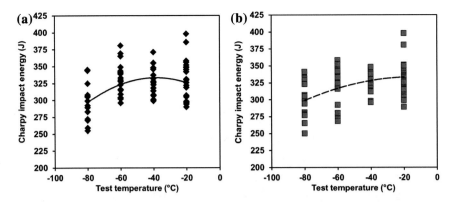

Fig. 7.2 Results of serial Charpy impact energy tests of the specimens from the X52MS-X56MS (**a**) and X60MS-X65MS (**b**) grade plates 20 mm thick

Table 7.8 Results of the HIC resistance test for the X52MS-X65MS grade steel plates 20.0 mm thick. (min–max)/average (Kholodnyi et al. 2016b)

Grade	HIC parameters (%)		
	CLR	CTR	CSR
X52MS	0–5.4 / 1.0	0–0.27 / 0.05	0–0.023 / 0.005
X56MS	0–11.3 / 2.1	0–0.53 / 0.08	0–0.067 / 0.009
X60MS	0–5.8 / 1.6	0–0.27 / 0.07	0–0.033 / 0.005
X65MS	0–10.7 / 0.9	0–0.23 / 0.02	0–0.067 / 0.004
Requirements	≤15	≤5	≤2

the rolled products met the requirements (Table 7.10). The ductile–brittle transition temperature of the experimental rolled steel was $T_{50} = -104$ °C. The experimental rolled products exhibited satisfactory HIC resistance (CLR = 4.2% and CTR = 0% at specified CLR ≤ 6% and CTR ≤ 1%) and SSC resistance upon testing cylindrical specimens (method A) at a load of 0.8 from 380 N/mm^2.

The rolled steel was used at the Volgorechensk Pipe Plant for the manufacture of gas–oil pipes of 159 mm in diameter. The pipes met the requirements for the mechanical characteristics of the K56 (X60) grade base metal and welded joints and for cracking resistance in hydrogen sulfide environments.

The X46 grade plates 20–22 mm thick were produced in the Rolling Mill 5000 of the Cherepovets Metallurgical Plant for the manufacture of electric-welded pipes of 720 mm in diameter for equipping the gas field containing hydrogen sulfide (Efron 2012). The chemical composition of the developed steel is given in Table 7.11.

Plates were rolled by controlled regimes at a finish rolling temperature in a range of 800–830 °C and subsequent accelerated cooling to a temperature of 560–580 °C. The plates had low content of non-metallic inclusions which corresponded to the

Table 7.9 Chemical composition of the steel (Zikeev et al. 2008)

Content of elements (wt%)													C_{eq}
C	Si	Mn	S	P	Al	Ti	Nb	V	Ca	N	H	
0.07	0.27	1.27	0.002	0.006	0.040	0.030	0.054	0.028	0.002	0.006	0.0002	0.31

Table 7.10 Mechanical properties of hydrogen sulfide-resistant skelp 6.0 mm thick from the K56 (X60) grade steel (Zikeev et al. 2008)

Mechanical properties	σ_T (N/mm^2)	σ_B (N/mm^2)	δ_5 (%)	KCV^{-60} (J/cm^2)	DWTT (%)
	495–520 508	615–635 626	27–27 27	148–169 158	100
Requirements	≥380	540–660	≥23	≥40	≥50

following numbers (estimation according to GOST 1778): a number of 0 for stitched oxides, ductile silicates, sulfides; a number of 1.0 for spot oxides; and a number of ≤2.0 for brittle silicates and non-deformable silicates. The plates met the specified requirements for HIC resistance (CLR ≤ 6% and CTR ≤ 1%) and SSC resistance ($\sigma_{thr} \geq 0.7 \cdot \sigma_T$).

The experience of the development of K50-K52 (X60) grade steel plates in the Rolling Mill 2800 at the Ural Steel Plant was considered in Pemov et al. (2013). Until 2007, the technology of controlled rolling followed by heat treatment according to the scheme of quenching with tempering was used for the production of hydrogen sulfide-resistant rolled steel. After the reconstruction in 2007–2009, which included the installation of a controlled cooling system after the finishing stand, it became possible to implement a more advanced controlled rolling technology with accelerated cooling. Table 7.12 presents the chemical composition of the K50-K52 (X60) grade steel for the manufacture of plates by both technologies.

The use of the CR + AC technology compared to the conventional technology with quenching and tempering made it possible to reduce the carbon, manganese, and niobium contents and to exclude the additions of nickel and vanadium. This led to an improvement in the weldability of steel. The plates 12.5–22.0 mm thick produced by both technologies met the requirements imposed on the K50-K52 (X60) grade steels for the hydrogen sulfide resistance of the C-2 group.

The "Ural Steel" Plant also implemented the technology of the batch production of K52 (X60) grade plates in a wide range of thicknesses from 9 to 36 mm. The plates are highly HIC resistant (CLR ≤ 3.9%. CTR ≤ 1.0%) and SSC resistant ($\sigma_{thr} \geq 0.7 \cdot \sigma_T$).

Thirty-year experience of the Europipe Company (Germany) on the development of technology and the manufacture of pipes resistant to cracking in H_2S-containing media was considered in Shabalov et al. (2017). It is noted that one of the most important factors of pipe resistance in sour environment is a strict control of the production process, from steelmaking and continuous casting (Fig. 7.3) to the manufacture of finished pipes.

Steelmaking provides economical alloying and minimum use of segregating elements. Low contents of carbon, sulfur, and phosphorus are of great importance. The desulfurization methods are used to reduce the sulfur content to ≤0.001%, and the calcium treatment is used to bind the remaining sulfur to non-deformable compounds. Soft reduction is used to decrease the central segregation during continuous casting.

Table 7.11 Chemical composition of the X46 grade steel (Efron 2012)

Elements (wt%)														C_{eq}	P_{cm}
C	Si	Mn	S	P	Cu	Ni	Cr	Al	Ti	Nb	Ca	N			
0.077	0.31	0.88	0.001	0.008	0.05	0.05	0.10	0.027	0.017	0.032	0.0021	0.006		0.26	0.14

Table 7.12 Chemical composition of the K50-K52 (X60) grade steel for the manufacture of plates in the hydrogen sulfide-resistant version (Pemov et al. 2013)

Technology	Content of elements (%) maximum or within the limits									C_{eq}	P_{cm}
	C	Si	Mn	S	P	Ti	Nb	V	Ni		
Quenching + tempering	0.11	0.35	1.20	0.003	0.010	0.020	0.080	0.050–0.060	0.20–0.30	0.36	0.18
CR + AC	0.09	0.30	1.10	0.003	0.010	0.020	0.050	–	–	0.31	0.16

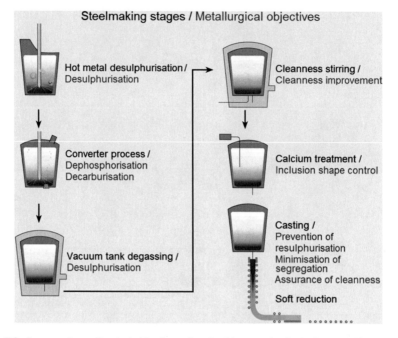

Fig. 7.3 Stages and metallurgical objectives of steelmaking practice for hydrogen sulfide-resistant steel by Europipe (Shabalov et al. 2017)

To obtain the required properties, the plates are produced by the technology of controlled rolling with accelerated cooling. The mechanical properties of steels for large diameter pipes especially strongly depend on the parameters of post-deformation cooling. A higher cooling rate beneficially affects the strength and toughness of the steel, which is especially important for plates of large thickness.

Europipe in 2006–2007 produced 430 thousand tons of pipes for operation in sour gas conditions. The X65 grade pipes had a diameter of 56″ and a wall thickness of 22.2–31.8 mm. Tables 7.13 and 7.14 present the chemical composition and mechanical properties of the pipe metal. The steels had low concentration of C < 0.05%, S < 0.0015%, and P < 0.015%, contained ≤1.48%Mn, and were microalloyed with Nb and Ti. For pipes of a larger wall thickness, the steel contained Ni and Cu additions. The base metal and the welded joint of the pipes had high impact toughness.

The HIC and SSC tests were performed in Solution A according to NACE TM0284 and NACE TM0177, respectively. The data given in Table 7.15 show that the base metal of the plates and the welded joint of the pipes have the required HIC resistance.

The improvement in the strength properties of pipes is an important factor in reducing the cost of pipelines. The X70 and X80 grade steels were not used until recently for this purpose, since they did not withstand the HIC test in Solution A. Taking into account the need for high-strength steels for sour gas service, the next logical step was the development of X70 and X80 grade pipes for severe sour gas

Table 7.13 Chemical composition of steels for X65 grade pipes of 56″ in diameter (Shabalov et al. 2017)

Wall thickness (mm)	Elements (wt%)					C_{eq}	P_{cm}
	C	Mn	S	P	Others		
22.2	<0.05	<1.46	<0.0015	<0.015	Nb, Ti	<0.33	<0.15
31.8	<0.05	<1.48	<0.0011	<0.015	Cu, Ni, Nb, Ti	<0.33	<0.15

Table 7.14 Mechanical properties (average) of the X65 grade pipes of 56″ in diameter (Shabalov et al. 2017)

Wall thickness (mm)	σ_T (N/mm^2)	σ_B (N/mm^2)	$\delta_{2''}$ (%)	$\sigma_{0.5}/\sigma_B$	DWTT^{-10} (%)	KV^{-30} (J)		
						Base metal	Fusion line	Weld metal
22.2	485	575	44	0.85	94	438	408	207
31.8	485	577	53	0.84	81	434	364	128

Table 7.15 Results of the HIC test of the base metal and weld joint of the X65 grade pipes of 56″ in diameter (Shabalov et al. 2017)

Wall thickness (mm)	CLR (%)	CTR (%)	CSR (%)
22.2	<5.3	<1.2	<0.6
31.8	<4.8	<1.5	<0.7
Requirements	<15	<5	<1.5

conditions. To determine the feasibility of their production, Europipe produced a pilot batch of X70 and X80 grade sour gas pipes of 20″ in diameter and 19.8 mm in wall thickness. The results of testing are presented in Tables 7.16, 7.17, and 7.18. The steels of both grades contained C < 0.05%, S < 0.0010%, P < 0.015%, and manganese > 1.50%. To provide the specified strength properties, the X70 grade steel contained Cu, Ni, Nb, and Ti, and the X80 grade steel was additionally alloyed with Mo.

The tests showed a high HIC resistance and the required mechanical properties of the X70 and X80 grade pipes. The steels exhibited a good SSC resistance upon four-point bend tests.

The X65MSO PSL2 grade pipes of 36″ in external diameter and 42.9 mm in a wall thickness are produced at the Welspun in Dahej, Gujarat, India. The manufacture of such pipes became possible due to the rational choice of the chemical composition of steel and the plate quality, strict control of the parameters of pipe forming according to the JCO process scheme and submerged arc welding.

Table 7.16 Chemical composition of steels for X70 and X80 grade pipes of 20″ in diameter and 19.8 mm in wall thickness (Shabalov et al. 2017)

Grade	Elements (wt%)					C_{eq}	P_{cm}
	C	Mn	S	P	Others		
X70	<0.05	>1.50	<0.0010	<0.015	Cu, Ni, Nb, Ti	0.34	0.15
X80	<0.05	>1.50	<0.0010	<0.015	Cu, Ni, Mo, Nb, Ti	0.37	0.17

Table 7.17 Mechanical properties (average) of X70 and X80 grade pipes of 20″ in diameter and 19.8 mm in wall thickness (Shabalov et al. 2017)

Grade	σ_T (N/mm²)	σ_B (N/mm²)	$\delta_{2''}$ (%)	$\sigma_{0.5}/\sigma_B$	DWTT^{-10} (%)	KV^{-30} (J)		
						Base metal	Fusion line	Weld metal
X70	500	585	45	0.84	100	337	251	209
X80	590	692	36	0.86	97.5	319	105	219

Table 7.18 Results of the HIC test of the base metal and weld joint of the X70 and X80 grade pipes of 20″ in diameter and 19.8 mm in wall thickness (Shabalov et al. 2017)

Grade	CLR (%)	CTR (%)	CSR (%)
X70	<3	0	0
X80	<6	<1	0
Requirements	<15	<5	<2

The pipes were manufactured from the plate 42.9 mm thick and 2712 mm wide produced at the VoestAlpine Grobblech GmbH, Austria. Chemical composition of the steel for the manufacture of the X65MSO grade plates is presented in Table 7.19. The steel has a low content of strongly segregating elements (C, S, and P) and 1.53% manganese, is economically microalloyed with Ti + V + Nb = 0.058%, and contains Cr = 0.19%. The weldability parameters of the steel are $C_{eq} = 0.33\%$ and $P_{cm} = 0.13\%$.

Figure 7.4 shows the technological stages of the plate production at the VoestAlpine Grobblech GmbH, Austria, plant. Continuously cast slab was rolled in a plate mill using thermomechanical processing technology followed by regulated accelerated cooling.

The rolled plates have a fine microstructure consisting of a matrix of quasipolygonal and polygonal ferrite with a high structure homogeneity across the plate thickness. The mechanical properties and HIC resistance of the plates fully meet the specified requirements (Tables 7.20 and 7.21).

Table 7.19 Chemical composition of the steel for the production of the X65MSO grade plates (Shabalov et al. 2017)

Elements (wt%)										C_{eq}	P_{cm}
C	Si	Mn	S	P	Ti + V + Nb	Cr	N	Al	Ca		
0.031	0.32	1.53	0.0006	0.007	0.058	0.19	0.004	0.033	0.001	0.33	0.13

(a)   **(b)**

(c)   **(d)**

Fig. 7.4 Technological scheme of steelmaking and plate rolling at VoestAlpine Grobblech GmbH, Austria: (**a**) steelmaking; (**b**) continuous casting; (**c**) thermomechanical rolling in a plate mill; and (**d**) accelerated cooling of the plate (Shabalov et al. 2017)

Table 7.20 Mechanical properties of the X65MSO grade plates 42.9 mm thick (min–max)/average (Shabalov et al. 2017)

Mechanical properties of plates	$\sigma_{0.5}$ (N/mm^2)	σ_B (N/mm^2)	$\delta_{2''}$ (%)	$\sigma_{0.5}/\sigma_B$	KV^{-23} (J)	DWTT^{-10} (%)	HV$_{10}$
	$\dfrac{465\text{–}508}{482}$	$\dfrac{556\text{–}573}{562}$	$\dfrac{32\text{–}34}{33}$	$\dfrac{0.83\text{–}0.89}{0.86}$	$\dfrac{438\text{–}452}{445}$	100%	$\dfrac{188\text{–}198}{193}$
Requirements	440–560	535–655	≥28	≤0.90	≥190	≥85	≤220

Table 7.21 Results of the HIC resistance test of the plates 42.9 mm thick from the X65MSO grade steel (min–max)/average (Shabalov et al. 2017)

HIC parameters (%)	CLR	CTR	CSR
	$\dfrac{0\text{–}0.54}{0.28}$	$\dfrac{0\text{–}0.03}{0.013}$	0
Requirements	≤10	≤3	≤1

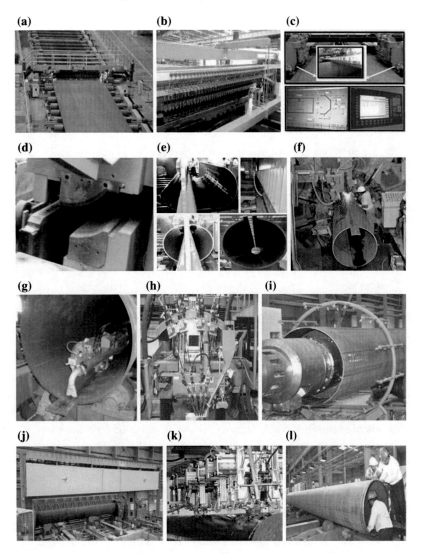

Fig. 7.5 Main technological stages of pipe processing from plate to pipe: (**a**) plate inspection; (**b**) plate ultrasonic testing; (**c**) edge milling; (**d**) edge crimping; (**e**) JCO; (**f**) continuous task welding; (**g**) welding of the internal seam; (**h**) welding of the outer seam; (**i**) mechanical expansion; (**j**) hydrostatic testing; (**k**) ultrasonic testing of the weld seam; and (**l**) final inspection (Shabalov et al. 2017)

Figure 7.5 shows the pipe production stages at the Welspun plant. Before the manufacture of the pipe, the plates are visually inspected and ultrasonically tested. At the next step, the longitudinal edges of the plate are trimmed by edge milling to achieve a predetermined width and then beveled to form a double "V" groove to accommodate the welding.

Table 7.22 Mechanical properties of the base metal of the X65MSO grade pipes of 36″ in diameter and 42.9 mm in wall thickness (min–max/average) (Shabalov et al. 2017)

Properties	$\sigma_{0.5}$ (N/mm^2)	σ_B (N/mm^2)	$\delta_{2″}$ (%)	$\sigma_{0.5}/\sigma_B$	KV^{-16} (J)	DWTT^{-17} (%)
	495–535 518	577–612 590	55–62 59	0.84–0.91 0.88	323–336 328	100%
Requirements	450–570	535–760	≥24	≤0.93	≥106	85

Table 7.23 Results of the Charpy impact test of the specimens taken from various locations of the X65MSO grade pipe of 36″ in diameter and 42.9 mm in wall thickness (min–max/average) (Shabalov et al. 2017)

Specimen location	Base metal		Heat-affected zone		Weld	
Specimen size	7.5 × 10 × 55 mm		10 × 10 × 55 mm		10 × 10 × 55 mm	
Temperature	−16 °C		−10 °C		−10 °C	
Charpy impact energy KV (J)	Unit value	Average	Unit value	Average	Unit value	Average
	319–338 328	323–336 328	326–467 426	367–447 426	108–213 160	122–201 160
Requirements	–	–	≥80	≥106	≥80	≥106

After this, the edges of the plate are bent between two stamps. Next, one longitudinal end of the plate is formed in gradual steps to a J-shape. The process is repeated from the other longitudinal end to form a C-shape. Finally, the pipe is converted into an O-shape.

After molding, continuous task welding is performed and then the outer and inner seams are welded. Expansion is carried out to a degree of deformation of 0.9–1.0%. After that, hydrostatic tests, ultrasonic testing, mechanical testing, and other inspection of pipe quality are performed.

Table 7.22 presents the mechanical properties of the base metal of the X65MSO grade pipes of 36″ in diameter and 42.9 mm in wall thickness. The average values of the yield strength and ultimate tensile strength increased by ~30 N/mm^2 due to plastic deformation upon JCO molding and mechanical expansion.

The difference in the values of the relative elongation is caused by different specimen shapes used upon tests of plates (round bar) and pipe (flat strip).

The hardness of the base metal, heat-affected zone (HAZ), and the weld seam was measured in accordance with the scheme shown in Fig. 7.6a. The average variations in hardness from the base metal to HAZ and the weld metal were 20 HV$_{10}$ and 40 HV$_{10}$, respectively (Fig. 7.6b).

The data given in Table 7.23 show the high Charpy impact energy of the base metal, heat-affected zone, and weld seam. The metal cut at 90°, 180°, and across the weld showed a high HIC resistance of the pipes (Table 7.24).

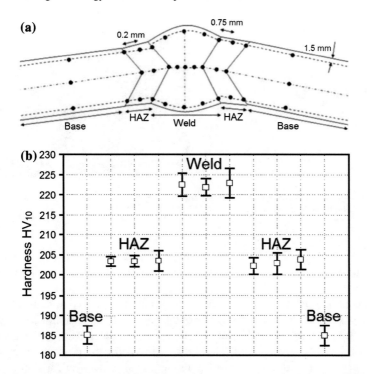

Fig. 7.6 Scheme (**a**) and the results (**b**) of the hardness measurements in the region of the welded seam of the X65MSO grade pipe of 36″ in diameter and 42.9 mm in wall thickness (Shabalov et al. 2017)

Table 7.24 Results of the HIC resistance test of the metal selected from various locations of the X65MSO grade pipe 36″ in diameter and 42.9 mm in wall thickness (Shabalov et al. 2017)

Specimen location	HIC parameters (%)		
	CLR	CTR	CSR
90° from weld	0–0.89 / 0.30	0	0
180° from weld	0	0	0
Across the weld	0–0.40 / 0.13	0	0
Requirements	≤10	≤3	≤2

Successful production of sour service pipes depends to a large extent on the continuously cast slab internal quality determined by a guaranteed cleanliness from nonmetallic inclusions and least possible centerline segregation and porosity. ArcelorMittal Lázaro Cárdenas (Mexico) has implemented key steelmaking and casting technologies for the production of sour service plates. The chemical composition of steels for the production of the X52 and X65 grade plates in this plant is presented in Table 7.25. For the reduction in the segregation heterogeneity of slabs, the steels

(a)   **(b)**

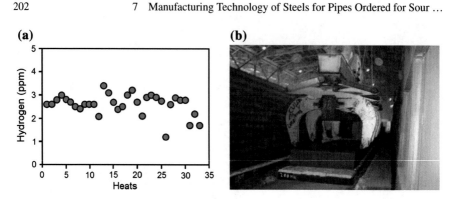

Fig. 7.7 Hydrogen content in heats before continuous casting (**a**) and stacking of slabs for slow cooling (**b**) (Shabalov et al. 2017)

have low contents of C, Mn, S, and P. The X65 grade steel of higher strength is characterized by increased contents of C, Mn, and Nb and contains additions of Cr, Cu, Mo, and V instead of Ni.

It is noted that the slab should have the minimum possible content of dissolved hydrogen gas for effective HIC resistance of the plates. Figure 7.7a shows the hydrogen content in the heats before continuous casting. It is seen that most heats have a hydrogen content of <3.0 ppm. The slabs intended for the production of plates for sour service pipes after casting are stored in a special chamber for slow cooling to allow a further diffusion of dissolved hydrogen from the slab center (Fig. 7.7b). Thus, the cooling rate of slabs can be decreased to 5–6 °C/h. The final hydrogen content in slabs is ≤2 ppm, which is preferred for effective HIC resistance in thick gauge linepipe plates.

The macrostructure of the cross section of a continuous slab 250 mm thick (Fig. 7.8a) indicates no apparent shrinkage cavities, dark spots, or centerline segregation band. Such microstructure can be qualified as a Mannesmann rating of class 1. Slabs were rolled to coils 6.25 mm thick and plates up to 20 mm thick. Figure 7.8b shows that the microstructure of the axial zone of the plate correlates with the axial zone of the slab and indicates the absence of any centerline segregation structural heterogeneity. The microstructure of the plates was examined for cleanliness from non-metallic inclusions. The observed inclusions have a globular shape, are fine, and remain non-deformed after rolling.

The tests for HIC and SSC resistance in Solution A of the X52 and X65 grade pipes made from hot-rolled coils and plates showed their high cracking resistance in hydrogen sulfide-containing media.

A significant reduction in the manganese segregation in the axial zone of the slab with a decrease in its total content in steel is shown in (Shabalov et al. 2017) Fig. 7.9. As the manganese content in steel is about 1.05 wt%, there are regions in the axial zone with a manganese concentration of 1.40%. While in the steel containing 0.30% manganese, no manganese segregation occurs in the axial zone. A decrease in man-

Table 7.25 Chemical composition of steels for the production of X52 and X65 grade skelp and plates (Shabalov et al. 2017)

Grade	Content of elements (wt%), maximum or within											
	C	Si	Mn	S	P	Al	Ti	Nb	N	Ca/S	Others
X52	0.02–0.04	0.25	1.00	0.001	0.012	0.025–0.045	0.020	0.050	0.003–0.006	2–4	Ni
X65	0.04–0.06	0.15	1.35	0.001	0.012	0.025–0.045	0.020	0.070	0.003–0.006	2–4	Cr, Cu, Mo, V

(a)   **(b)**

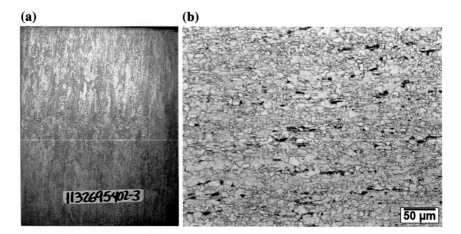

Fig. 7.8 Macrostructure of the continuously cast slab (**a**) and microstructure of the axial zone of the plate from the X65 grade hydrogen-resistant steel (**b**) (Shabalov et al. 2017)

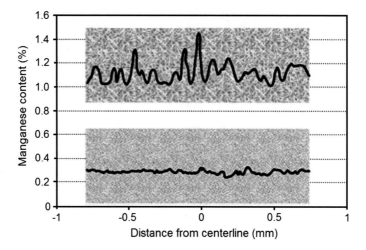

Fig. 7.9 Distribution of manganese concentration in the axial zone of slabs from steels with different manganese contents (Shabalov et al. 2017)

ganese segregation also minimizes the central segregation of carbon and phosphorus and the formation of manganese sulfides in the axial zone of the slab.

A batch of coils 12.70 and 14.27 mm thick was made from trial heats with a reduced Mn content of 0.24–0.29%. The chemical compositions of the steels are shown in Table 7.26. The steels had low concentrations of C = 0.035–0.040%, S = 0.0016–0.0024%, and P = 0.012–0.013%.

Because of low manganese content, alloying with 0.40–0.41%Cr, 0.084–0.086%Nb, Cu, and Ni was used for additional hardening. The steel for the production of coil 14.27 mm thick additionally contained 0.052%V.

Table 7.26 Chemical compositions of low-manganese steels for the production of coils 12.70 and 14.27 mm thick (Shabalov et al. 2017)

Thickness (mm)	Elements (wt%)											C_{eq}
	C	Si	Mn	S	P	Al	Cr	Nb	V	Others	
12.70	0.040	0.17	0.29	0.0024	0.013	0.040	0.40	0.086	0.001	Cu, Ni, Ti, Ca	0.20
14.27	0.035	0.18	0.24	0.0016	0.012	0.034	0.41	0.084	0.052		0.20

Table 7.27 Mechanical properties of the X52 grade coils 12.70 and 14.27 mm thick (Shabalov et al. 2017)

Thickness (mm)	σ_T (N/mm²)	σ_B (N/mm²)	δ (%)	σ_T/σ_B	KV^0 (J)	$DWTT^0$ (%)	HV_{10}
12.70	432	481	39	0.90	232	>95	189
14.27	448	498	34	0.90	219	>95	192

Table 7.28 Chemical compositions of steels for the manufacture of X80 grade strips 12 and 18 mm thick (Shabalov et al. 2017)

Content of elements (wt%)							
C	Si	Mn	P + S	Cu + Ni	Nb + Ti + V	Ca	Others
≤0.07	≤0.50	≤1.80	≤0.005	≤0.80	≤0.20	>0.002	Cr, Mo

Table 7.29 Mechanical properties of H_2S-resistant X80 grade strip 12 and 18 mm thick (Shabalov et al. 2017)

Thickness (mm)	σ_T (N/mm²)	σ_B (N/mm²)	δ (%)	σ_T/σ_B	KV^{-5} (J)	HV_{10}
12	582	633	35	0.92	341	196
18	561	653	40	0.86	485	189

The mechanical properties of the skelp met the requirements for the X52 grade steels (Table 7.27). The test results showed its good HIC resistance: CLR = 0%, CTR = 0%, and CSR = 0%.

For the manufacture of hydrogen sulfide-resistant pipes, it is necessary to take into account the possible changes in both mechanical properties and cracking resistance in H_2S-containing media from plate to pipe as a result of deformation during pipe production. The effect of pipe manufacture on the properties of hydrogen sulfide-resistant pipes of 20″ in diameter from strips 12 and 18 mm thick was shown in POSCO (Shabalov et al. 2017). The chemical composition and the mechanical properties of the X80 grade rolled products are presented in Tables 7.28 and 7.29. The coils had a high HIC resistance (CAR = 0%) and SSC resistance upon four-point bending test.

To evaluate the effect of pipe processing, tensile tests were performed on specimens taken from the pipe at the 180° position from the welding line. For the pipes of 12 and 18 mm in wall thicknesses, the t/D ratios were 0.0236 and 0.0354, respectively. A higher t/D ratio indicates a higher deformation of metal upon pipe forming. Figure 7.10 shows the effect of the t/D ratio on the change in the yield strength from coil to pipe.

For a pipe of 12 mm in wall thickness, the yield strength decreased by ~50 N/mm² relative to that of the strip, while the yield strength of the pipe of 18 mm in wall thickness increased by about 30 N/mm². This change in the yield strength indicates that a higher t/D ratio of 0.0354 causes an increase in the yield strength due to strain

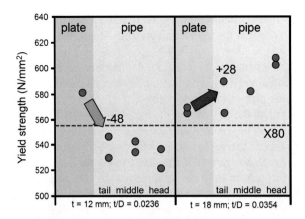

Fig. 7.10 Change in yield strength upon the manufacture of pipes of 20″ in diameter and 12 mm ($t/D = 0.0236$) and 18 mm ($t/D = 0.0354$) in wall thickness (Shabalov et al. 2017)

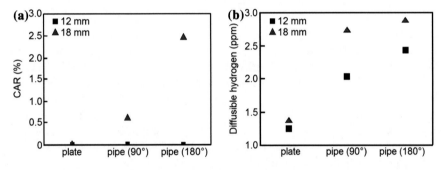

Fig. 7.11 Results of the HIC test (**a**) and the amount of absorbed hydrogen (**b**) measured after HIC test of the plates 12 and 18 mm thick and pipes of 20″ in diameter made from the plates (Shabalov et al. 2017)

hardening upon the forming process, and a lower t/D ratio of 0.0236 results in a decreased yield strength because of the Bauschinger effect. The yield strength of the weld was over 560 N/mm^2 for both 12 and 18 mm pipes.

The ultimate tensile strength values for plates and pipes including weld metal were at the same level and exceeded 625 N/mm^2. As a result of pipe forming from X80 grade plates, the strength of the pipe of 12 mm in wall thickness decreased to X70 grade, while the pipes of 18 mm in wall thickness remained at the X80 grade level.

The results of the HIC tests of plates and pipes are shown in Fig. 7.11a. It is seen that the plates 12 and 18 mm thick are not susceptible to HIC. The test of the pipe of 12 mm in wall thickness also showed a high HIC resistance. On the contrary, test of the pipe of 18 mm in wall thickness revealed a tendency to form hydrogen cracks. For the pipe metal located at 90° from weld, CAR was 0.61%, while at 180° CAR was 2.46%, i.e., was higher by a factor of four.

The amount of hydrogen absorbed by the steel was measured after the HIC test (Fig. 7.11b). It is seen that the hydrogen concentration in the pipe was higher than in the plate. The sample taken from the 180° position had a higher hydrogen content than the sample taken from the 90° position. At the same time, the metal of the pipe with $t = 18$ mm contained a higher amount of hydrogen captured upon the HIC test than the metal of the pipes with $t = 12$ mm.

The data obtained on the tendency of the metal pipes to HIC and on the amount of absorbed hydrogen correlate well with each other. Based on the results of the study, one can conclude that an increase in the degree of deformation expressed through an increase in the t/D ratio as well as a higher degree of deformation realized upon pipe forming at the 180° position from the weld reduce the HIC resistance of the pipes.

Numerous patents of leading Chinese, Russian, Japanese, and South Korean metallurgical companies manufacturing plates and large diameter pipes for sour gas application contain data on technological parameters and specific features of the production of pipe steels, their chemical composition, HIC resistance, and sulfide stress cracking, as well as mechanical properties.

In most of the patents, the properties of steel correspond to the K48-K52 and X52-X70 grades. The carbon content, as a rule, is in a range of 0.02–0.08%. The manganese concentration in the steels varies widely. The specified properties of the steels are achieved due to alloying with chromium, nickel, copper, and molybdenum, as well as due to additions of microalloying elements such as titanium, niobium, and vanadium.

The sulfur content is strictly limited by all companies. At the same time, not only the calcium content, but also the ratio between sulfur and calcium contents is specified. Some steels contain REM. A number of patents limit the maximum contents of oxygen, nitrogen, and hydrogen.

To achieve high properties of pipe steels, including high HIC and SSC resistances, high strength, cold resistance, weldability, it is not enough to choose only the optimal chemical composition of steel. It is necessary to optimize the production methods such as steelmaking, hot plastic deformation, and cooling.

An important condition for the achievement of high specified properties is the formation of fine structure with the regulated number and size of structure constituents (ferrite, pearlite, bainite), fine carbonitride particles precipitated under certain rolling conditions, and limited quantity of non-metallic inclusions.

References

Barykov, A. B. (Ed.). (2016). *Development of steel manufacturing technology for steel rolled product and pipe in the Vyksa production area: Coll. works*. Moscow: Metallurgizdat.

Efron, L. I. (2012). Metal science in big metallurgy, pipe steels. Moscow: Metallurgizdat.

Golovanov, A. V., Zikeev, V. N., Kharchevnikov, V. P., et al. (2005). Composition and production conditions for cold-resistant and H_2S-resistant rolled strip for oil and gas pipe. *Steel in Translation, 35*(9), 60–62.

Kholodnyi, A. A., Matrosov, Y. I., Matrosov, M. Y., & Sosin, S. V. (2016a). Effect of carbon and manganese on low-carbon pipe steel hydrogen-induced cracking resistance. *Metallurgist, 60*(1), 54–60.

Kholodnyi, A. A., Sosin, S. V., Matrosov, Y. I., & Karmazin, A. V. (2016b). Development of the production technology at the ISW "Azovstal" of sheets for hydrogen sulfide-resistant pipes of large diameter of the X52-X65 strength classes. *Problems of Ferrous Metallurgy and Materials Science, 4,* 26–34.

Kholodnyi, A. A., Matrosov, Y. I., & Sosin, S. V. (2017). Influence of molybdenum on microstructure, mechanical properties and resistance to hydrogen induced cracking of plates from pipe steels. *Metallurgist, 61*(3), 230–237.

Kudashov, D. V., Mursenkov, E. S., Stepanov, P. P., et al. (2017). Assimilation of pipe steel extra-furnace treatment and casting technology with specification for resistance to H_2S media under casting and rolling complex conditions. *Metallurgist, 61*(8), 656–665.

Matrosov, Y. I., Kholodnyi, A. A., Matrosov, M. Y., et al. (2015). Effect of accelerated cooling parameters on microstructure and hydrogen cracking resistance of low-alloy pipe steels. *Metallurgist, 59*(1), 60–68.

Pemov I. F., Nizhel'skii D. V., Naumenko A. A., et al. (2013) Production of corrosion-resistant strip (strength class K50–K52) at OAO Ural'skaya Stal. *Steel in Translation, 43*(4):215–220.

Shabalov, I. P., Matrosov, Yu I, Kholodnyi, A. A., et al. (2017). *Steel for gas and oil pipelines resistant to fracture in hydrogen sulphide-containing media.* Moscow: Metallurgizdat.

Zikeev, V. N., Kharchevnikov, V. P., Filatov, N. V., & Anuchin, K. V. (2008). Developing and testing structural steels with superior resistance to cold and to cracking in hydrogen-sulfide-bearing media for use in the production of electric-welded (HFC) oil-field tubing. *Metallurgist, 52*(9), 491–497.

Printed in the United States
By Bookmasters